Intrinsic
Bioremediation

BIOREMEDIATION

The *Bioremediation* series contains collections of articles derived from many of the presentations made at the First, Second, and Third International In Situ and On-Site Bioreclamation Symposia, which were held in 1991, 1993, and 1995 in San Diego, California.

First International In Situ and On-Site Bioreclamation Symposium

1(1)   *On-Site Bioreclamation: Processes for Xenobiotic and Hydrocarbon Treatment*

1(2)   *In Situ Bioreclamation: Applications and Investigations for Hydrocarbon and Contaminated Site Remediation*

Second International In Situ and On-Site Bioreclamation Symposium

2(1)   *Bioremediation of Chlorinated and Polycyclic Aromatic Hydrocarbon Compounds*

2(2)   *Hydrocarbon Bioremediation*

2(3)   *Applied Biotechnology for Site Remediation*

2(4)   *Emerging Technology for Bioremediation of Metals*

2(5)   *Air Sparging for Site Bioremediation*

Third International In Situ and On-Site Bioreclamation Symposium

3(1)   *Intrinsic Bioremediation*

3(2)   *In Situ Aeration: Air Sparging, Bioventing, and Related Remediation Processes*

3(3)   *Bioaugmentation for Site Remediation*

3(4)   *Bioremediation of Chlorinated Solvents*

3(5)   *Monitoring and Verification of Bioremediation*

3(6)   *Applied Bioremediation of Petroleum Hydrocarbons*

3(7)   *Bioremediation of Recalcitrant Organics*

3(8)   *Microbial Processes for Bioremediation*

3(9)   *Biological Unit Processes for Hazardous Waste Treatment*

3(10) *Bioremediation of Inorganics*

Bioremediation Series Cumulative Indices: 1991-1995

For information about ordering books in the Bioremediation series, contact Battelle Press. Telephone: 800-451-3543 or 614-424-6393. Fax: 614-424-3819. Internet: sheldric@battelle.org.

Intrinsic Bioremediation

Edited by

Robert E. Hinchee
Battelle Memorial Institute

John T. Wilson
U.S. EPA Ground Water Protection and Remediation Division

Douglas C. Downey
Parsons Engineering Science, Inc.

BATTELLE PRESS
Columbus • Richland

Library of Congress Cataloging-in-Publication Data

Hinchee, Robert E.
    Intrinsic bioremediation / edited by Robert E. Hinchee, John T. Wilson,
    Douglas C. Downey.
        p.   cm.
    Includes bibliographical references and index.
    ISBN 1-57477-002-0 (hc : acid-free paper)
    1. Bioremediation—Congresses. 2. Nature—Congresses. I. Hinchee,
    Robert E. II. Wilson, John T. III. Downey, Douglas C.
    TD192.5.I58   1995
    628.5′2—dc20   95-32246
                                                                        CIP

Printed in the United States of America

Additional copies may be ordered through:
Battelle Press
505 King Avenue
Columbus, Ohio 43201, USA
1-614-424-6393 or 1-800-451-3543
Fax: 1-614-424-3819
Internet: sheldric@battelle.org

CONTENTS

FOREWORD

This book and its companion volumes (see overleaf) comprise a collection of papers derived from the Third International In Situ and On-Site Bioreclamation Symposium, held in San Diego, California, in April 1995. The 375 papers that appear in these volumes are those that were accepted after peer review. The editors believe that this collection is the most comprehensive and up-to-date work available in the field of bioremediation.

Significant advances have been made in bioremediation since the First and Second Symposia were held in 1991 and 1993. Bioremediation as a whole remains a rapidly advancing field, and new technologies continue to emerge. As the industry matures, the emphasis for some technologies shifts to application and refinement of proven methods, whereas the emphasis for emerging technologies moves from the laboratory to the field. For example, many technologies that can be applied to sites contaminated with petroleum hydrocarbons are now commercially available and have been applied to thousands of sites. In contrast, there are as yet no commercial technologies commonly used to remediate most recalcitrant compounds. The articles in these volumes report on field and laboratory research conducted both to develop promising new technologies and to improve existing technologies for remediation of a wide spectrum of compounds.

Intrinsic remediation can be defined as the combined effect of natural destructive and nondestructive processes to reduce a contaminant's mobility, mass, and associated risks. Nondestructive mechanisms include sorption, dilution, and volatilization. These processes often serve to reduce the concentration and toxicity of contaminants; this reduction may enhance natural processes that destroy contaminants. The primary emphasis of this volume is destructive attenuation processes, specifically aerobic and anaerobic biodegradation.

Intrinsic aerobic biodegradation of normal aliphatics and simple aromatics is well documented as a means of remediating soil and groundwater contaminated with fuel hydrocarbons; Brown et al. (pp. 77-84) state that intrinsic aerobic degradation should be considered an integral part of the remediation process. However, several papers presented in this volume provide geochemical evidence that dissolved hydrocarbons at fuel-contaminated sites are biodegraded anaerobically, as well as aerobically. In fact, anaerobic processes may accomplish greater mass reduction than do aerobic processes. Wiedemeier et al. (pp. 31-51), Wilson et al. (pp. 91-100), and Toze et al. (pp. 117-125) show that denitrification, iron reduction, sulfate reduction, and methanogenesis can provide significant anaerobic pathways for dissolved fuel hydrocarbon biodegradation. The predominant electron acceptors for different zones of the aquifer appear to be dictated by the availability of individual electron acceptors and localized redox conditions at a given site.

Just as destruction of fuel hydrocarbons by intrinsic processes has been demonstrated in numerous field studies, there is growing evidence that natural processes significantly influence more recalcitrant chemicals, such as polycyclic aromatic hydrocarbons (PAHs), chlorinated organics, and mixed hydrocarbons.

Studies by Ginn et al. (pp. 153-162) and King et al. (pp. 171-179) deal with intrinsic processes that immobilize and biodegrade PAH compounds. The Danish researchers Heron et al. (pp. 143-151) and Rügge et al. (pp. 127-133) have observed geochemical patterns indicative of anaerobic biodegradation in aquifers contaminated with mixed landfill wastes. Significant research is underway internationally to better understand, quantify, and predict how intrinsic remediation processes attenuate chlorinated organics, specifically dissolved organic solvents such as trichloroethylene (TCE) and perchloroethylene (PCE).

Lee et al. (pp. 205-222) describe attenuation of highly concentrated chlorinated solvents in an industrial landfill. Based on modeling studies, they estimated that volatile chlorinated organics were attenuating at a rate of 70% per year. If the residence time of a plume is about 10 years, rates of this magnitude will reduce contamination a thousandfold. Cox et al. (pp. 223-231) describe the degradation of trichloroethene and 1,1,1-trichloroethane in groundwater contaminated by a septic waste lagoon. Near the source in anaerobic groundwater, TCE was dechlorinated to 1,2-dichloroethene and vinyl chloride, and 1,1,1-dichloroethane was dechlorinated to 1,1-dichloroethane and chloroethane. As the contaminated groundwater moved away from the source, it became oxygenated, allowing the aerobic biodegradation of chloroethane and vinyl chloride. This combination of anaerobic and aerobic treatment probably occurs in most plumes of chlorinated solvents. Intrinsic remediation of chlorinated organics frequently is incomplete. However, a considerable component of the environmental hazard is removed by intrinsic processes. Innovative remedial strategies will include the intrinsic processes as part of the total remedial effort.

The editors would like to recognize the substantial contribution of the peer reviewers who read and provided written comments to the authors of the draft articles that were considered for this volume. Thoughtful, insightful review is crucial for the production of a high-quality technical publication. The peer reviewers for this volume were:

Abdul Abdul, *General Motors*
C. P. Antworth, *U.S. Air Force*
Sabine Apitz, *NRAD*
Daniel J. Arp, *Oregon State University*
Jeffrey Barbaro, *University of Waterloo* (Canada)
James Barker, *University of Waterloo* (Canada)
Michael A. Beckman, *B.P. Barber and Associates, Inc.*
Ralph E. Beeman, *DuPont Co.*
Robert C. Borden, *North Carolina State University*
Robin L. Brigmon, *Oak Ridge Inst. for Science and Education*
Fred Brockman, *Battelle Pacific Northwest*
Kim Broholm, *Technical University of Denmark*
Francis H. Chapelle, *U.S. Geological Survey, WRD*
Chen Yu Chiang, *Shell Development Co.*
Chris Du Plessis, *University of Natal* (Rep. of South Africa)

R. Ryan Dupont, *Utah State University*
Neal D. Durant, *Johns Hopkins University*
Christopher Finton, *Battelle Columbus*
D.R.J. Grootjen, *DSM Research* (The Netherlands)
Dick B. Janssen, *University of Groningen* (The Netherlands)
Michael Jawson, *U.S. Environmental Protection Agency*
Ursula Jenal-Wanner, *Stanford University*
Leslie Karr, *U.S. Navy*
Marian W. Kemblowski, *Utah State University*
Rajagopal Krishnamoorthy, *Groundwater Technology, Inc.*
Bruce Labelle, *CA Dept. of Toxic Substances Control*
Paul M. McAllister, *Shell Development Co.*
Scott Miller, *AGRA Earth & Environmental, Inc.*
Mercedes Mondecar, *Clark Atlanta University*
Craig Payne, *Battelle Columbus*
John W. Ratz, *Parsons Engineering Science, Inc.*
Hanadi S. Rifai, *Rice University*
Guy W. Sewell, *U.S. Environmental Protection Agency*
Manish Shah, *Battelle Pacific Northwest*
Robert Sharp, *Montana State University*
Stephen H. Shoemaker, *DuPont Co.*
Judith L. Sims, *Utah State University*
Thomas Stauffer, *U.S. Air Force*
Mark Trudell, *Alberta Research Council*
Michael J. Truex, *Battelle Pacific Northwest*
Todd H. Wiedemeier, *Parsons Engineering Science, Inc.*
Paula J. Woolcar, *Lancaster University* (UK)

Finally, we want to recognize the key members of the production staff, who put forth significant effort in assembling this book and its companion volumes. Carol Young, the Symposium Administrator, was responsible for the administrative effort necessary to produce the ten volumes. She was assisted by Gina Melaragno, who tracked draft manuscripts through the review process and generated much of the correspondence with the authors, co-editors, and peer reviewers. Lynn Copley-Graves oversaw text editing and directed the layout of the book, compilation of the keyword indices, and production of the camera-ready copy. She was assisted by technical editors Bea Weaver and Ann Elliot. Loretta Bahn was responsible for text processing and worked many long hours incorporating editors' revisions, laying out the camera-ready pages and figures, and maintaining the keyword list. She was assisted by Sherry Galford and Cleta Richey; additional support was provided by Susan Vianna and her staff at Fishergate, Inc. Darlene Whyte and Mike Steve proofread the final copy. Judy Ward, Gina Melaragno, Bonnie Snodgrass, and Carol Young carried out final production tasks. Karl Nehring, who served as Symposium Administrator in 1991 and 1993, provided valuable insight and advice.

The symposium was sponsored by Battelle Memorial Institute with support from many organizations. The following organizations cosponsored or otherwise supported the Third Symposium.

Ajou University–College of Engineering (Korea)
American Petroleum Institute
Asian Institute of Technology (Thailand)
Biotreatment News
Castalia
ENEA (Italy)
Environment Canada
Environmental Protection
Gas Research Institute
Groundwater Technology, Inc.
Institut Français du Pétrole
Mitsubishi Corporation
OHM Remediation Services Corporation
Parsons Engineering Science, Inc.
RIVM–National Institute of Public Health and the Environment
    (The Netherlands)
The Japan Research Institute, Limited
Umweltbundesamt (Germany)
U.S. Air Force Armstrong Laboratory–Environics Directorate
U.S. Air Force Center for Environmental Excellence
U.S. Department of Energy Office of Technology Development
    (OTD)
U.S. Environmental Protection Agency
U.S. Naval Facilities Engineering Services Center
Western Region Hazardous Substance Research Center–
    Stanford and Oregon State Universities

Neither Battelle nor the cosponsoring or supporting organizations reviewed this book, and their support for the Symposium should not be construed as an endorsement of the book's content. Rob Hinchee conducted the final review and selection of all papers published in this volume, making use of the essential input provided by the peer reviewers and other editors. He takes responsibility for any errors or omissions in the final publication.

John Wilson
Doug Downey
Rob Hinchee
June 1995

Intrinsic Bioattenuation for Subsurface Restoration

*Hanadi S. Rifai, Robert C. Borden,
John T. Wilson, and C. Herb Ward*

ABSTRACT

Intrinsic bioattenuation has recently evolved as a viable remediation alternative at a number of sites where the risk of exposure to contaminants is within acceptable standards. Important mechanisms controlling the instrinsic bioattenuation include advection, dispersion, sorption, dissolution from a residual source, and abiotic and biological transformations. Because intrinsic bioattenuation is a plume management strategy, it requires characterizing and monitoring these processes. Intrinsic bioattenuation involves an assessment of risks to public health and the environment, and consequently requires prediction of the fate and transport of contaminants at the candidate sites. This paper reviews the processes controlling intrinsic bioremediation and summarizes case histories in which intrinsic bioattenuation has been observed at sites contaminated with petroleum hydrocarbons and chlorinated solvents. The key steps in evaluating natural attenuation as a remedial alternative are summarized.

INTRODUCTION

In the absence of human intervention, many contaminant plumes will develop until they reach a quasi-steady-state condition. At steady state, the contaminant plume is no longer growing in extent and may shrink somewhat over time. Major processes controlling the size of the steady-state plume include (1) release of dissolved contaminants from the source area, (2) downgradient transport of the contaminants and mixing with uncontaminated groundwater, (3) volatilization, and (4) abiotic and biologically mediated transformations of the contaminants of concern.

Intrinsic bioattenuation is a plume management strategy by which the natural assimilation processes are monitored and used to limit adverse impacts of groundwater contamination. This strategy also requires an assessment of risks to public health and other environmental receptors. A successful implementation of intrinsic bioattenuation at a field site requires adequate site hydrogeological, chemical, and biological characterization; detailed data analysis to determine whether contaminants are being attenuated and/or removed from the aquifer; modeling of the fate and transport of the dissolved groundwater plume; and, finally, long-term monitoring to confirm and ensure protection of human health and the environment.

PROCESSES CONTROLLING THE STEADY-STATE CONTAMINANT DISTRIBUTION

Physical

The primary physical processes affecting the distribution of contaminants in groundwater include advection, dispersion, sorption, volatilization, and dissolution from residual contaminants located in the source area. Advection is the process by which contaminants are transported with the flow of groundwater. Dispersion accounts for mechanical and molecular mixing processes. Both advection and dispersion reduce contaminant concentrations but do not cause a net loss of mass of contaminants in the aquifer. Higher advection and dispersion cause more spreading and more dilution of a dissolved contaminant plume.

Sorption describes the partitioning of contaminants between the aqueous phase and the solid aquifer matrix. Sorptive processes tend to reduce the dissolved contaminant concentrations and limit the migration of the aqueous-phase plume, but they do not result in a loss of contaminant mass from the aquifer. Under steady-state conditions, sorption will not affect the final contaminant distribution; however, sorption will delay the development of a steady-state plume.

Volatilization refers to the partitioning of a contaminant between the aqueous phase in the saturated zone and the vapor phase in the unsaturated zone. While volatilization actually removes mass from the aquifer, it is not thought to be a significant attenuation mechanism except in situations where the groundwater table is less than 15 ft (4.6 m) deep and the unsaturated zone consists of relatively transmissive soils. Chiang et al. (1989) estimated that volatilization resulted in a mass loss of benzene of less than 5% at a gas plant facility in Michigan.

Dissolution from residual contaminants located in the source area is by far the most significant physical process that controls the extent of a contaminant plume. In the case of petroleum hydrocarbons, contaminants may dissolve from a lens of mobile hydrocarbons floating on the water table, or from residual hydrocarbons trapped in the soil matrix above and/or below the water table. Seasonal fluctuations in the water table can cause additional "smearing" and dissolution from residual source areas. Until sources are depleted, a contaminant plume will expand until it reaches a quasi steady state.

Abiotic and Biologically Mediated Transformations

Aerobic and anaerobic biodegradation processes are believed to account for both contaminant concentration reduction and loss of pollutant mass from the aquifer. Abiotic transformations such as hydrolysis and dehydrohalogenation also attenuate concentrations and contaminant mass in an aquifer, but they are significant only for specific chemicals such as chlorinated solvents.

Aerobic biodegradation relies on dissolved oxygen as the electron acceptor used by the microorganisms. Petroleum hydrocarbons are generally very amenable to aerobic biodegradation in aquifers with dissolved oxygen concentrations exceeding 1 to 2 mg/L. Many shallow-water-table aquifers contain background dissolved oxygen concentrations between 1 to 12 mg/L depending on the temperature of the groundwater. While aerobic biodegradation takes place at relatively higher rates than anaerobic processes, it is often limited by the available supply of oxygen to a contaminant plume. Once the background dissolved oxygen is consumed in the center of the plume, aerobic biodegradation is limited to the edges of the contaminant plume, where the dissolved contaminants come into contact with oxygen-rich groundwaters.

Anaerobic processes refer to a variety of biodegradation mechanisms that use NO_3^-, SO_4^{2-}, Fe^{3+}, and CO_2 as terminal electron acceptors. Anaerobic biodegradation dominates the interior of a contaminant plume. Both petroleum hydrocarbons and chlorinated solvents are believed to biodegrade to varying degrees under anaerobic conditions. However, the rates of biodegradation are often slower than under aerobic conditions.

Hydrocarbon Biodegradation. Most organic compounds found in crude, refined oil and fuels are known to degrade under aerobic conditions. The aerobic biodegradation of benzene, toluene, ethylbenzene, xylenes, naphthalene, methyl-napthalenes, dibenzofuran, and fluorene has been confirmed by a large number of laboratory and field studies. Aerobic processes are relatively fast and limited by the rate at which oxygen is supplied to a contaminant plume. Because of this phenomenon, Rifai et al. (1988) modeled aerobic biodegradation as an instantaneous reaction between oxygen and the hydrocarbons.

Current research efforts have also shown that monoaromatic compounds degrade under anaerobic conditions. This biodegradation occurs with NO_3^- (Evans et al. 1991), Fe^{3+} (Lovley and Lonergan 1990), SO_4^{2-} (Edwards et al. 1991), and carbon dioxide (Grbic-Galic and Vogel 1987; Wilson et al. 1986) as electron acceptors.

Benzene has been found to be recalcitrant to anaerobic biodegradation in laboratory studies using nitrate and sulfate as electron acceptors (Kuhn and Suflita 1989a,b; Edwards et al. 1991). However, some laboratory and field studies demonstrated the degradation of all monoaromatic hydrocarbons under denitrifying, sulfate-reducing, and methanogenic conditions (Major et al. 1988; Cozzarelli et al. 1990; Vogel and Grbic-Galic 1986; Barker and Wilson 1992; Wilson et al. 1994a). The aromatic compounds may be oxidized first to phenols or organic acids, then transformed to volatile fatty acids before complete mineralization. Anaerobic biodegradation of aromatic hydrocarbons is therefore associated with the production of fatty acids, methane, carbon dioxide, solubilization of iron, and reduction of nitrate and sulfate.

Adaptation of an aromatic plume to nitrate as a terminal electron acceptor seems to occur readily once oxygen is depleted. Natural biodegradation through nitrate respiration should be similar to oxidative biodegradation. Depending on its concentration, sulfate can also be an important electron acceptor. Acton and Barker (1992) demonstrated the sulfate reduction of toluene and *m*-xylene in a forced-gradient injection experiment at an active landfill in Ontario, Canada. Benzene, *o*-xylene, ethylbenzene, and 1, 2, 4- trimethylbenzene were not degraded at the site.

Chlorinated Aliphatic Hydrocarbons (CAH) Transformation. Chlorinated solvents, consisting primarily of CAHs, can be transformed by chemical and biological processes to form a variety of other CAHs (McCarty, 1994). Reduction of tetrachloroethene (PCE) and trichloroethene (TCE) to ethene has occurred at many sites, although transformations often are not complete. Freedman and Gossett (1989) provided evidence for the conversion of PCE and TCE to ethene and de Bruin et al. (1992) reported complete reduction to ethane. McCarty (1994) lists the possible transformations for a number of the predominant chlorinated solvents in groundwater (Table 1). McCarty (1994) indicates that methanogenesis is the most favorable mechanism for complete reduction of PCE and TCE to ethene.

A number of researchers have confirmed the biological transformation of CAHs at field sites. Major et al. (1991) reported evidence for bioattenuation of PCE to ethene and ethane at a chemical transfer facility in North Toronto. Fiorenza et al. (1994) presented data on the chemical and biological transformation of trichloroethane (TCA) to 1,1-dichoroethane (1, 1-DCA) and that of PCE and TCE to *cis*-dichloroethene (*c*-DCE), vinyl chloride (VC), and ethene at a manufacturing plant in Ontario. Beck (1994) reported the degradation of 1, 1, 1-TCA, PCE, and TCE to ethene and methane at the Dover Air Force Base (AFB) in Delaware. McCarty and Wilson (1992), Haston et al. (1994), Kitanidis et al. (1993), McCarty et al. (1991), and Wilson et al. (1994b) confirmed the

TABLE 1. Environmental conditions generally associated with reductive transformations of chlorinated solvents.

Chlorinated Solvent	Redox Environment			
	All	Denitrification	Sulfate Reduction	Methanogenesis
Carbon tetrachloride		CT → CF	CT → CO_2+Cl	
1,1,1-Trichloroethane	TCA → 1,1-DCE +CH_3COOH		TCA → 1,1-DCA	TCA → CO_2+Cl
Tetrachloroethylene			PCE → 1,2-DCE	PCE → ethene
Trichloroethylene			TCE → 1,2-DCE	TCE → ethene

intrinsic biodegradation of chlorinated solvents at the St. Joseph, Michigan, Superfund site.

In addition to biological transformations, chemical transformations of some CAHs can occur in groundwater through elimination or hydrolysis. TCA is one of the main chlorinated solvents that can be transformed chemically in ground-water under all conditions likely to be found and within a reasonable time frame (McCarty 1994). The rate of chemical transformations is usually expressed using a first-order reaction. TCA chemical transformation, for example, leads to the formation of 1,1-DCE and acetic acid with a reported average half-life of less than 1 year at a temperature of 20°C.

CASE STUDY—INTRINSIC BIOREMEDIATION OF A UST RELEASE

The underground storage tank (UST) release in Rocky Point, North Carolina, provides a representative example of a dissolved benzene, toluene, ethyl benzene, and xylenes (BTEX) plume undergoing intrinsic biodegradation using oxygen, nitrate, iron, and sulfate as terminal electron acceptors (Borden et al. 1995). The water table aquifer consists of mostly fine-grained, dark-gray or greenish-gray, micaceous, glauconitic, slightly silty, and compact quartz sand. The sand appears to be very homogeneous throughout the site with only a few exceptions. This sand is overlain by lower-permeability clays and clayey sands that form a surface confining layer throughout the site. The average groundwater velocity is approximately 30 m/y. The organic-carbon content of the sand is relatively low (0.1%), and consequently sorption is not a major attenuation mechanism for the moderately soluble BTEX fraction.

Spatial Distribution of BTEX and Indicator Parameters

Background groundwater contains moderate levels of dissolved oxygen (2 to 3 mg/L), nitrate (1 to 6 mg/L as N), and sulfate (20 to 30 mg/L). Background dissolved iron is low (< 0.5 mg/L), and the groundwater is acidic (pH < 5) with low buffering capacity (alkalinity ~ 6 mg/L as $CaCO_3$) and low levels of dissolved CO_2 (15 to 30 mg/L as C). At the upgradient edge of the BTEX plume, residual hydrocarbon is trapped below the water table in the sand aquifer. As uncontaminated groundwater enters this region, soluble hydrocarbons partition out of the nonaqueous-phase (NAPL) and into the aqueous phase. Figure 1 shows the observed variation in BTEX components, electron acceptors, and indicator parameters in a profile along the dissolved-hydrocarbon-plume centerline. Several distinct zones can be identified where different oxidation-reduction processes dominate.

At the upgradient edge of the BTEX plume, a portion of the soluble hydro-carbons released from the residual NAPL are immediately degraded using oxygen and nitrate carried into this zone by the flowing groundwater. Dissolved iron increases to 29 mg/L because of reduction of insoluble iron

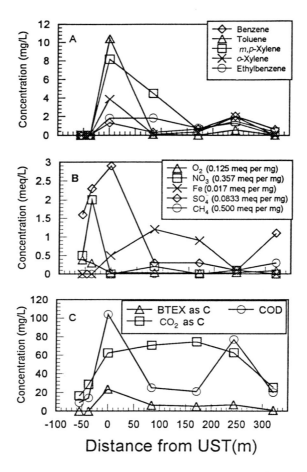

FIGURE 1. **Variations in (A) BTEX components, (B) electron acceptors, and (C) indicator parameters in a profile along the plume centerline.**

oxides associated with the sediment. In this region, the dominant electron acceptor is nitrate, followed by iron and oxygen. Dissolved CO_2 increases from 16 to 60 mg/L as carbon (C) because of oxidation of organic matter. The pH also rises from ~ 4.7 to 5.8 because of consumption of H^+ during iron reduction.

During transport downgradient from the source, toluene and o-xylene decline rapidly followed by m, p-xylene and benzene. Ethylbenzene does not decline notably with distance. This pattern is apparently due to preferential biodegradation of the o-xylene isomer (and toluene) by subsurface microorganisms. Sulfate decreases from 34 to 0.7 mg/L, total sulfur decreases from 37 to 5 mg/L as SO_4, and dissolved iron increases from 29 to 65 mg/L. The large decline in sulfate and increase in dissolved iron indicate that both sulfate and iron reduction are occurring. The smaller decline in total sulfur suggests that sulfate is being reduced but is not being removed as ferrous sulfide (FeS) precipitates. Ion chromatographic analysis of the groundwater indicates that thiosulfate is a major component of the nonsulfate sulfur. Oxygen and nitrate are not significant electron acceptors in this portion of the plume since they were already consumed during transport through the source area.

The dissolved hydrocarbon plume becomes slightly narrower with distance downgradient. The limited spreading of the plume is apparently due to the combined effects of anaerobic biodegradation in the plume center and aerobic biodegradation at the sides of the plume. The background dissolved oxygen concentration varies from 2 to 3 mg/L, whereas in the center of the plume dissolved oxygen is below the field detection limit (0.3 mg/L). As the plume spreads in width by dispersion, oxygen in the uncontaminated groundwater mixes with BTEX, enhancing biodegradation at the plume sides. The zone of highest BTEX concentration also moves vertically downward with increasing distance downgradient. The vertical drop along the length of the plume is due to surficial groundwater recharge that both adds a layer of clean, uncontaminated water on top of the plume and enhances biodegradation by introducing oxygen and other electron acceptors with the recharge water.

Rates of Intrinsic Bioremediation

Field implementation of intrinsic bioremediation requires an accurate estimation of in situ biodegradation rates. Three different approaches were applied at the Rocky Point site to estimate intrinsic bioremediation rates: (1) comparing peak contaminant concentrations in monitoring wells versus travel time from the source (Borden et al. 1994); (2) monitoring laboratory microcosms under ambient (anaerobic) aquifer conditions (Hunt et al. 1995); and (3) monitoring in situ test chambers to determine compound loss over time (Hunt et al. 1995). The in situ test chambers were installed approximately midway down the plume in the iron-reducing zone, at the same location used to collect the sediment for the laboratory microcosms. Results from each of these approaches are compared in Table 2.

In the laboratory microcosms, a distinct order of biodegradation was observed. Toluene and *o*-xylene appeared to be the most biodegradable followed by *m*-, *p*-xylene and benzene, with ethylbenzene being the least biodegradable. This same order of disappearance is also seen in the field data. Unfortunately, the rates of biodegradation estimated from the field data are vastly different from the laboratory data. In the laboratory microcosms, the benzene, toluene, and xylene isomers were degraded from between 2,000 and 3,000 μg/L each to below detection limit in 400 days. Ethylbenzene was only slightly degraded over this period. If similar biodegradation rates were observed in the field, the benzene, toluene, and xylene plumes would completely biodegrade over 100 to 200 ft (30 to 60 m). Yet significant concentrations of these dissolved hydrocarbons persist over 1,300 ft (396 m) downgradient from the source. The cause of this discrepancy is not well understood.

TABLE 2. Comparison of intrinsic bioremediation rates from field monitoring, laboratory microcosms and in situ test chambers.

Compound	Field Rate (d⁻¹)	Laboratory Rate (d⁻¹)	In Situ Rate (d⁻¹)
Benzene	0.0002	0.024	0.004
Toluene	0.0021	0.045	0.012
Ethylbenzene	0.0015	0.002	N.S.[a]
o-Xylene	0.0021	0.056	N.S.
m-,p-Xylene	0.0013	0.02[b]	0.014

(a) Not significant at 95% level.
(b) Only *m*-xylene in laboratory microcosms.

Figures 2a and 2b show the vertical distribution of benzene, ethylbenzene, *m*-, *p*-xylene, and two trimethylbenzene isomers (mesitylene and pseudocumene) in a multilevel sampler located 800 ft (244 m) downgradient from the source near the location of the in situ test columns. The concentrations of toluene and *o*-xylene were too low to be shown on these figures. In both figures, the vertical distribution of the more recalcitrant compounds (benzene, ethylbenzene, and mesitylene) is relatively consistent. In contrast, there are large changes in the concentration of the more biodegradable compounds (*m*-, *p*-xylene and pseudocumene). This suggests that the rate of biodegradation may be significantly different between adjoining layers.

The observed differences in field and laboratory biodegradation rates could be due to changes in the activity of different layers against the pollutants. In the field, if one layer is not active against the pollutants, the vertically averaged concentration measured with long-screened wells would remain high and the apparent biodegradation rate would be low. In contrast, within in situ test columns and laboratory microcosms, groundwater is forced into contact with sediment of differing activities. This would result in a higher apparent biodegradation rate.

CASE STUDY—INTRINSIC BIOREMEDIATION OF TCE IN GROUND WATER

The groundwater at the St. Joseph, Michigan, site is contaminated with CAHs at concentrations ranging from 10 to 100 mg/L. The contaminants are divided into eastern and western plumes as the suspected sources were situated over a groundwater divide. Both plumes contain TCE; *cis*- and *trans*-1, 2-dichloroethene (*c*-DCE and *t*-DCE), 1, 1,-dichloroethene (1, 1-DCE), and VC. Significant levels of ethene and methane were measured at the site (Kitanidis

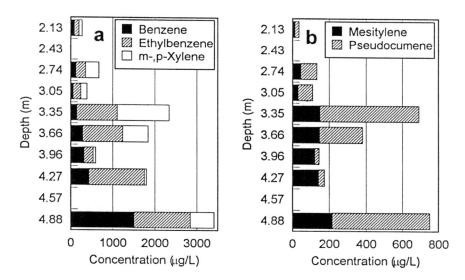

FIGURE 2. Vertical distribution of dissolved hydrocarbons in iron-reducing zone 170 m downgradient from the former UST.

et al. 1993; McCarty and Wilson 1992), confirming the natural attenuation of TCE.

McCarty and Wilson (1992) delineated contours of the chemical oxygen demand (COD)—a surrogate for the capacity of a donor to supply electrons— and correlated them with contours of chlorinated aliphatic compounds (see Figure 3a and b). The authors found a correlation between COD decrease and transformations of TCE to VC. Also shown in Figure 3a are the locations of three transects where data were collected for the detailed bioattenuation characterization. Table 3 summarizes the electrons released by TCE reduction to different products, along with the equivalent amount of COD decrease. Essentially, the reduction of 1 mole or 131 g of TCE to ethene releases six electrons, and an equivalent decrease of 48 g of COD is needed.

Table 4 summarizes the CAHs, ethene, and methane found at some of the monitoring locations. The average concentration of methane at depths of 25 m or more was 6 mg/L. This concentration of methane corresponds to an equivalent COD decrease of 24 mg/L. However, McCarty and Wilson (1992) measured COD rease, as high as 200 mg/L. They attribute this inconsistency to either dilution effects between the lagoon and the detailed characterization location, or to the presence of other electron acceptors such as nitrate and sulfate.

Apparent Degradation Constants

Wilson et al. (1994b) studied the western plume at the site to estimate the contaminant mass flux, and to estimate apparent degradation constants. Data

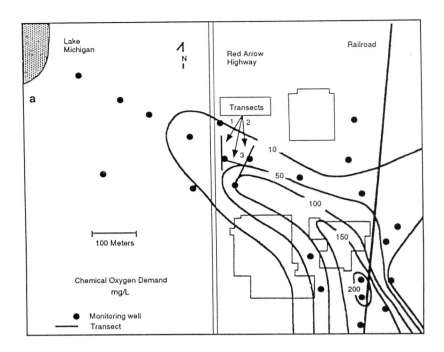

FIGURE 3a. COD contours at the St. Joseph, Michigan, NPL site.

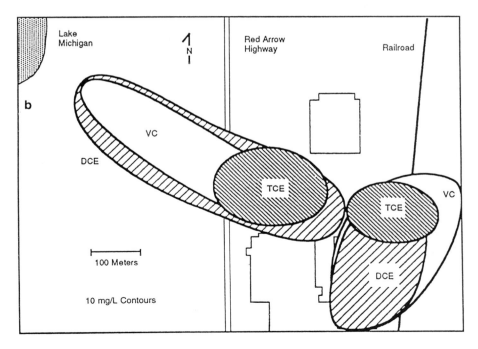

FIGURE 3b. CAH (10 mg/L) contours at the St. Joseph, Michigan, NPL site.

TABLE 3. Half-reactions indicating electron equivalents of change and associated equivalent COD decrease associated with change.

Transform product	Mol. Wt.	Half-Reaction	Equiv. COD decrease g COD/g product[a]
Methane	16	$CO_2+8H^++8e^- \rightarrow CH_4+2H_2O$	4
DCE	97	$CHCl=CCl_2+H^++2e^- \rightarrow CHCl=CHCl+Cl^-$	0.16
VC	62.5	$CHCl=CCl_2+2H^++4e^- \rightarrow CH_2=CHCl+2Cl^-$	0.51
Ethene	28	$CHCl=CCl_2+3H^++6e^- \rightarrow CH_2=CH_2+3Cl^-$	1.71

(a) From $(O_2+4H^+4e^- \rightarrow 2H_2O)$, the half-reaction for oxygen, one electron equivalent of COD equals one-fourth mole of molecular oxygen or 8 grams, thus equiv. COD decrease = (8n/Mol. Wt), where n is the number of electrons in the half-reaction.

TABLE 4. Concentrations of CAHs, ethene, and methane found at selected sampling locations along detailed characterization transects and the equivalent COD decrease associated with the products.

Sample[a]		TCE	1,1-DCE	c-DCE	t-DCE	VC	Ethene	CH$_4$	Total
1-2-70	mg/L	4.03	0.09	4.14	0.81	3.19	6.62	4.61	
	Equiv. COD								
	Decrease, mg/L	0.0	0.01	0.66	0.13	0.41	11.32	18.44	30.97
	% of Equiv. COD	0.0	0.0	2.1	0.4	1.3	36.6	59.5	
1-3-75	mg/L	12.80	0.27	16.90	0.67	56.40	2.25	6.62	
	Equiv. COD								
	Decrease, mg/L	0.0	0.04	2.70	0.11	28.80	3.84	26.00	61.5
	% of Equiv. COD	0.0	0.1	4.4	0.2	46.8	6.2	42.2	
22-1-75	mg/L	0.44	0.09	13.40	0.21	1.46	3.15	7.43	
	Equiv. COD								
	Decrease, mg/L	0.0	0.01	2.14	0.03	0.74	5.39	29.72	38.03
	% of Equiv. COD	0.0	0.0	5.6	0.1	1.9	14.2	78.1	
2-6-65	mg/L	0.51		4.70	0.03	2.66	4.27	11.72	
	Equiv. COD								
	Decrease, mg/L	0.0	0.0	0.75	0.0	1.36	5.98	46.88	54.97
	% of Equiv. COD	0.0	0.0	1.4	0.0	2.5	10.9	85.3	
3-2-80	mg/L	2.53	0.01	0.90	0.03	0.27	4.87	11.59	
	Equiv. COD								
	Decrease, mg/L	0.0	0.0	0.14	0.0	0.14	8.32	46.36	54.96
	% of Equiv. COD	0.0	0.0	0.3	0.0	0.3	15.1	84.4	

(a) First value is transect number, second value is borehole number, and third value is depth of sample below ground surface in feet.

collected in 1991 from three transects near the source of the western plume and data collected in 1992 from two additional transects were used in the analysis (see Figure 4 for locations of the transects). The mass estimates combined with the flow velocities were used to estimate the mass flux at each transect. As

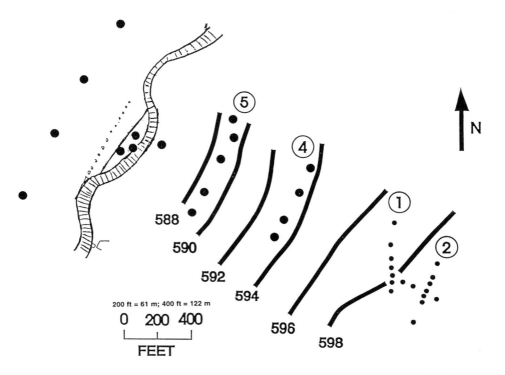

FIGURE 4. **Locations of the 1991 and 1992 sampling transects at the St. Joseph, Michigan, NPL site.**

would be expected, the mass fluxes decline toward the downgradient edge of the plume.

The mass per unit thickness of TCE at transects 2, 4, and 5 was used to estimate first-order degradation constants. Table 5 lists the computed apparent loss coefficients for three different estimates of the hydraulic conductivity at the site. The rate for TCE degradation ranges from 0.0048 to 0.011 wks^{-1} between the upgradient transects 2 and 4. This rate increases up to 0.023 wks^{-1} between transects 4 and Lake Michigan.

DEMONSTRATING INTRINSIC BIOREMEDIATION IN THE FIELD

Initial Monitoring to Determine Feasibility of Intrinsic Bioremediation

Extensive field monitoring is initially conducted to determine if intrinsic bioremediation is feasible as a remedial alternative at a site. Rifai (1990) and Borden (1990) proposed preliminary sampling protocols for soils, groundwater,

TABLE 5. Apparent loss coefficients at St. Joseph, Michigan.

Compound	Low Conductivity Estimate (1wk)	Avg. Conductivity Estimate (1 wk)	High Conductivity Estimate (1 wk)
Transects 2 to 4			
TCE	0.0048	0.0074	0.011
c-DCE	0.0064	0.0097	0.013
t-DCE	0.0076	0.012	0.0016
1,1-DCE	0.0066	0.010	0.0135
VC	0.0023	0.0035	0.0047
Transects 4 to 5			
TCE	0.016	0.025	0.033
c-DCE	0.010	0.016	0.021
t-DCE	0.010	0.016	0.021
1,1-DCE	0.012	0.018	0.024
VC	0.011	0.017	0.023
Transects 5 to Lake			
TCE	0.011	0.018	0.023
c-DCE	0.038	0.059	0.079
t-DCE	0.0092	0.014	0.019
1,1-DCE	—	—	—
VC	0.053	0.081	0.11

and soil gas (see Table 6). These protocols were intended for UST sites and mostly focused on defining the electron acceptor distribution in the ground-water, and on quantifying the by-products of biodegradation in groundwater and soil gas. Since then, the Air Force Center for Environmental Excellence (AFCEE) has developed a detailed protocol for site characterization in support of the natural attenuation alternative at Air Force facilities (Wiedemeier et al. 1994).

At a minimum, site characterization in an intrinsic bioattenuation protocol should provide data on the location and extent of contaminant sources, the extent and distribution of dissolved contaminants, groundwater geochemical data (i.e., concentrations of electron acceptors and by-products of biodegradation mechanisms), geologic characterization data, and hydrogeologic parameters such as hydraulic conductivity, gradients, and potential migration pathways.

The site characterization data are analyzed to quantify the extent of intrinsic bioattenuation. Overall, three indicators of natural attenuation can be developed from this information:

TABLE 6. Proposed parameters for field measurements.

Medium	Parameter
Soil	Microbial counts/activity
	TOC[a]
	BTEX
Groundwater	Temperature
	Dissolved oxygen
	Carbon dioxide
	Conductance
	pH, TDS[b]
	Redox potential
	Ca, Mg, Na, Mn, Fe, SO_4, Cl
	Total alkalinity
	BOD[c], COD[d], TOC, BTEX
	NO_3, NO_2, NH_4, PO_4
Soil Vapor	O_2, CO_2, CH_4, H_2S

(a) TOC = total organic carbon;
(b) TDS = total dissolved solids;
(c)   BOD = biological oxygen demand;
(d) COD = chemical oxygen demand.

1. Compound disappearance: one of the most convincing arguments for natural attenuation involves demonstrating the disappearance of a dissolved
   organic chemical at the site relative to the persistence of another "conservative" or "recalcitrant" organic (internal standard). In some cases, it is sufficient to demonstrate that the extent of migration of the organic of concern
   is less than that of the "conservative" tracer, and thus its transport downgradient is being limited by natural attenuation. In cases where compound
   disappearance cannot be related to an internal standard, it may be possible
   to demonstrate mass loss as a function of time for the organic of interest.
   Another alternative involves analyzing the peak concentrations of the
   organic at the different monitoring wells downgradient from the source.
   This analysis should demonstrate the overall decline of these concentrations as a function of time and distance.
2. Loss of electron donors: measuring dissolved oxygen concentrations and
   those of other electron acceptors can provide a key indicator of natural
   attenuation. Reduced oxygen, nitrate, and sulfate concentrations within
   the plume relative to their background concentrations are considered to be
   strong evidence of intrinsic bioattenuation. Many field studies, for instance,
   have correlated depressed oxygen concentrations within the center of a
   plume with high concentrations of the dissolved contaminant.

3. Degradation products: the accumulation of dissolved iron and the production of carbon dioxide, hydrogen sulfide, and methane are additional indicators of biological attenuation at intrinsic bioremediation sites.

One of the interesting questions that is currently being investigated by researchers is whether it may be possible to complete a mass balance on the supply of electron donors and electron acceptors at a given field site. This question is complicated, because of sampling and field-data collection limitations. Other complicating factors involve the temporal nature of the distributions of the electron acceptors and donors.

Risk Assessment

The successful application of the natural attenuation alternative involves an exposure- and risk-assessment analysis that considers the location of receptors at a site in relation to the extent of the contaminant plume. While a detailed discussion of risk assessment is beyond the scope of this paper, it is important to note that intrinsic bioremediation will only be viable as a remedial alternative if it can be demonstrated that little risk to human health and the environment will be incurred as a consequence of this management strategy. To demonstrate the viability of intrinsic bioremediation, it may be necessary to conduct fate and transport predictions of the future conditions at contaminated sites.

Prediction of Plume Migration—Modeling Approaches

Two key questions need to be answered when determining the viability of natural attenuation as a remediation alternative, namely (1) how far the dissolved plume will migrate before it is attenuated to below a predetermined cleanup standard, and (2) how long it will take for the attenuation process to "clean up" the plume. Both questions can be readily answered using analytical and numerical models of fate and transport. Analytical models are simpler to use, but they are limited in their capabilities to simplified hydrogeologic scenarios. Numerical models are more complicated, but can be used to simulate heterogeneous systems and more complex hydrogeologic and contaminant scenarios.

Numerous fate and transport models have been developed over the years. The majority of these models simulate advection, dispersion, sorption, and some form of source representation. A smaller number of these models, however, can actually simulate complex biological and chemical transformation processes. In the analytical modeling arena, the most common method for simulating biodegradation is through the use of a first-order decay coefficient. The contaminant of concern is assumed to biodegrade exponentially, and the modeler specifies the first-order decay constant for a given site and a given contaminant. This leaves the modeler with the dilemma of selecting a first-order decay constant.

While many laboratory and field studies have developed first-order decay constants for a variety of contaminants under a number of hydrogeologic scenarios, these constants are not readily transferable to other sites. Another problem noted by Rifai (1994) is that the first-order decay model does not account for electron acceptor limitations and thus can overestimate the effect of biodegradation on a given system. Connor et al. (1994) proposed using an instantaneous reaction expression similar to that used in BIOPLUME II (Rifai et al. 1988) as an alternative to the first-order decay model. Additionally, Rifai and Hopkins (1995) have developed, through modeling with BIOPLUME II, "electron-acceptor" limited decay coefficients for scenarios with contaminant sources removed and continuous contaminant sources (Tables 7a and 7b) to provide more "applicable" decay constants.

Finally, and for sites where a long history of monitoring exists, it may be possible to estimate a first-order decay constant based on the observed mass loss in the aquifer. The dissolved concentrations for different sampling events are used to generate an estimate of the dissolved mass in the aquifer as a function of time. The resulting data allow the modeler to estimate a first-order decay constant. It should be mentioned, however, that this procedure is highly dependent on the density of the sampling network and would not be very accurate for sites with a limited number of monitoring wells.

Numerical models, as mentioned earlier, can provide more simulation capabilities than analytical models. The BIOPLUME II model (Rifai et al. 1988) allows the user to simulate a heterogeneous aquifer system with a variable flow field. The BIOPLUME II model is one of the few two-dimensional models that can simulate the transport of an electron acceptor (oxygen in this case) and its reactions with the aquifer contaminants. The model currently simulates the instantaneous reaction between oxygen and aromatic hydrocarbons. Rifai et al.

TABLE 7a. Aerobic decay rates for removed source scenario. [a]

	Gradient	12 mg/L	9 mg/L	6 mg/L	3 mg/L
Sand	0.01	3.8E-04	2.5E-04	1.8E-04	1.0E-04
	0.001	1.0E-06	1.0E-06	7.5E-07	7.5E-07
Silt	0.01	1.0E-05	1.0E-05	7.5E-06	5.0E-06
Clay	0.01	1.0E-05	1.0E.05	1.0E-06	1.0E-06

(a) The decay rates listed in this table were obtained through BIOPLUME II model simulations. A continuous source scenario assuming 100 mg/L was injected at a rate of 10 gal/d (38 L/d) in the sand hydrogeologic environment. A lower injection rate, 1 gal/d (3.8 L/d), was used in the silt and clay hydrogeologic environment to minimize mounding in the model. Centerline concentrations at different receptor well locations modeled for oxygen-limited degradation were matched to concentrations using first-order decay.

TABLE 7b. Aerobic decay rates for continuous source scenario. [b]

	Gradient	12 mg/L	9 mg/L	6 mg/L	3 mg/L
Sand	0.01	–3.2E-01	–2.1E-01	–1.3E-01	–6.8E-02
	0.001	–3.3E-02	–2.4E-02	–1.7E-02	–1.0E-02
	0.0001	–8.7E-03	–6.3E-03	–4.1E-03	–2.5E-03
Silt	0.01	–5.8E-03	–4.1E-03	–3.0E-03	–1.6E-03
	0.001	–2.8E-04	–2.4E-04	–2.1E-04	–1.8E-04
	0.0001	–2.5E-05	–2.4E-04	–2.1E-05	–2.3E-05
Clay	0.01	–4.0E-04	–3.4E-04	–3.0E-04	–2.6E-04
	0.001	–2.1E-04	–3.5E-05	–3.0E-05	–2.7E-05
	0.0001	–4.3E-06	–3.7E-06	–3.2E-06	–2.9E-06

(b) The decay rates listed in this table were obtained through BIOPLUME II model simulations. A removed source scenario assuming a plume 45 ft (14 m) and 50 ft
(15 m) long with an initial concentration of 10 mg/L was simulated. Log of present
mass dissolved at 1 month to 5 to 10 years was plotted over time. The decay rate
is the slope of the plotted line.

(1995) are extending the BIOPLUME II model to allow the simulation of multiple electron acceptors within a contaminant plume. A number of other numerical biodegradation models exist in addition to BIOPLUME II (Table 8). The
majority of these models, however, are either one dimensional or of a propri-
etary nature.

One of the difficulties encountered in using numerical models is determining what data are required and how to incorporate the field data into the
modeling process. Most fate and transport models require an estimate of the
aquifer thickness, matrix conductivity, porosity, and sorptive characteristics.
Additionally, most models require some description of the hydraulic and
hydrologic stresses on the system in the form of boundary conditions, or
recharge and discharge specifications. One of the most complicated parameters to estimate for numerical models is the source representation because, in
most cases, the history of contamination at the site is not known with any
degree of certainty. Finally, biodegradation models require input on the electron acceptor availability within the aquifer.

The process of simulating natural attenuation at a site using a numerical
model requires (1) calibrating the model to the hydraulics at the site so that the
model can emulate the direction of flow and observed groundwater velocities
in the field; and (2) calibrating the model to simulate existing contamination
conditions. Once those two steps have been completed, the numerical model
can be used to determine the distribution of contaminants at the site as a function of time.

TABLE 8. Biodegradation models (from Bedient et al. 1994).

Name	Description	Author(s)
—	1-D, aerobic, microcolony, Monod	Molz et al. (1986)
BIOPLUME	1-D, aerobic, Monod	Borden and Bedient (1986)
—	1-D, analytical first order	Domenico (1987)
BIO1D	1-D, aerobic and anaerobic, Monod	Srinivasan and Mercer (1988)
—	1-D, cometabolic, Monod	Semprini and McCarty (1991)
—	1-D, aerobic anaerobic, nutrient limitations, microcolony, Monod	
—	1-D, aerobic, cometabolic, multiple substrates, fermentative, Monod	Celia et al. (1989)
BIOPLUME II	2-D, aerobic, instantaneous	Rifai et al. (1988)
—	2-D, Monod	MacQuarrie et al. (1990)
BIOPLUS	2-D, aerobic, Monod	Wheeler et al. (1987)
ULTRA	2-D, first order	Tucker et al. (1986)
—	2-D, denitrification	Kinzelbach et al. (1991)
—	2-D, Monod, biofilm	Odencrantz et al. (1990)

One of the problems faced in modeling intrinsic bioattenuation of organic chemicals at sites is the fact that the observed data usually incorporate the effects of advection, dispersion, sorption and biodegradation. Therefore, it may be difficult to estimate the advective and dispersive components independently from these data. A possible solution is to use a "conservative tracer" or an "internal standard" in the calibration of the numerical model that does not sorb or biodegrade. For example, when simulating gasoline spills, it may be possible to use methyl terbutyl ether (MTBE) concentrations in the calibration process if the data exist. (MTBE does not sorb or biodegrade and thus would reflect the advective and dispersive characteristics of the aquifer). The biodegradation and sorption of the aromatic hydrocarbons within the gasoline plume can be readily estimated by comparing the BTEX plume to the MTBE plume. Wiedemeier et al. (1995a,b) have also suggested the use of tetramethyl-benzene as a more recalcitrant internal standard at fuel-spill sites.

SUMMARY OF INTRINSIC BIOREMEDIATION FIELD SITES

Over the past decade, a large number of sites undergoing intrinsic bioremediation have been studied in detail. Some of the major characteristics of selected sites are provided in Table 9. Upgradient background concentrations of oxygen, nitrate, and sulfate provide an indication of the concentrations potentially available for biodegradation. Elevated concentrations of Fe^{2+} and CH_4 in the plume reflect the importance of iron reduction and methanogenic fermentation. At many sites, significant concentrations of iron, nitrate, and sulfate are available to support hydrocarbon biodegradation. Although methane

TABLE 9. Characteristics of intrinsic bioremediation field sites.

Site	Aquifer Type	pH	Background O₂ (mg/L)	NO₃-N (mg/L)	SO₄ (mg/L)	Plume Fe²⁺ (mg/L)	CH₄ (mg/L)
Borden, Ontario (Barker et al. 1987)	glaciolacustrine, medium to fine sand						
Borden, Ontario (Barbaro et al. 1992)	glaciolacustrine, medium to fine sand	7.4	<0.07	<0.05	220		
Rocky Point, NC (Borden et al. 1995)	marine, fine sand	4.7	2.5	6	30	102	0.4
Kalkaska, MI (Chiang et al. 1989)	medium-coarse sand, gravel interbeds	5.5-7.0	9	0.8	1-6		
Columbus, MS (MacIntyre et al. 1993)	fluvial, hetero- geneous sands and clays		2.6-3.8				
Sleeping Bear, MI (Wilson et al. 1994a)	glacial outwash, coarse sand/gravel	6.4	2.4	15	20	17-28	24-30
Eglin AFB, FL (Wilson et al. 1994a)	sands and silty peats	5.6-6.7				8.0	17
Hill AFB-1, UT (Wiedemeier et al. 1994)	thin channel sands in deltaic deposits	~7	6	8	100	50	2
Patrick AFB, FL (Wiedemeier et al. 1994)	fine to coarse marine sand with shell fragments	~7	3.7	0.3	86	1.9	14.6
Fairfax, VA (Buscheck et al. 1993)		5	2.5-4			43	
Sampson Co., NC (Borden personal communication 1995)	clayey and silty sands	4-5	8	17	8	0.4	<0.01
Traverse City (Wilson et al. 1990)	glaciolacustrine, medium sand with gravel	7	9	3	10	12	17
Broward Co., FL (Caldwell et al. 1992)		7.2	2-3	0.3	<5	1.8	
Pensacola, FL (Godsy et al. 1992; Bekins et al. 1993)	poorly sorted fine to coarse deltaic sands	6.9	0.04	2	6	33	13
Bemidji, MN (Baedecker et al. 1993)	glacial outwash, coarse sand with silt layers	7.6	7.7	0.001	2	16	12
Galloway, NJ (Cozzarelli and Baedecker 1992)	fine to coarse sand, perched water table on clay lens	4.7	6	9	24	37	0.007
Perth, Australia (Thierrin et al. 1993)	medium to fine aeolian sand	5.8	0.4	0.1	20-100	1.4	
Manufacturing Plant (Davis et al. 1994)	glacial silty sand over bedrock	6.9	1.5	<0.005			
Cliffs-Dow (Klecka et al. 1990)	coarse sand and gravels	6.1	1.2	1.7	48		
Hill AFB-2, UT (Dupont et al. 1994)			4	6.9			

fermentation has been documented at several sites, it appears to be less important than iron and sulfate reduction.

Biodegradation results from field and laboratory studies are reported in Table 10. Laboratory results are reported only if the laboratory study was designed to simulate field conditions. Effective decay rates have been estimated in the field using several approaches. The most reliable approach is to calculate a mass balance for a known mass of contaminant injected into an aquifer, using a dense network of monitoring points. This approach is feasible only when a pulse of contaminant is injected. For continuous sources at steady state, the degradation rates may be calculated from the change in total mass flux across several lines of monitoring wells. Both the mass balance and mass flux approaches are expensive to implement because of the high number of monitoring points required. At most field sites, the only feasible approach is to estimate the degradation rate from a plot of peak contaminant concentration versus travel time from the source. This general approach has been modified by normalizing the contaminant concentrations to an internal standard that is poorly biodegradable, has similar sorption and volatilization properties as the contaminant, and is present in the waste source. Ideally, the internal-standard approach should correct for changes in concentration due to dilution.

There is a wide range in reported biodegradation rates. Reported first-order degradation rates for benzene range from nondetectable to approximately 1% per day, with an average of approximately 0.2% per day. Degradation rates for other hydrocarbons are typically somewhat higher, but in the same general range. Higher biodegradation rates occur most frequently at sites containing higher concentrations of sulfate in the background water and higher concentrations of dissolved iron in the plume. Where high concentrations of methane are observed, biodegradation rates are often lower.

SUMMARY

Past research has shown that intrinsic bioremediation can control the migration of dissolved hydrocarbon plumes. Field biodegradation rates are often lower than would be expected based on laboratory results, but are often sufficient to contain the contaminant plumes within reasonable transport distances. Fewer data are available on chlorinated hydrocarbon plumes, but ongoing studies suggest that intrinsic bioremediation may also be technically feasible at these sites. At many sites intrinsic bioremediation alone may be the best alternative available for risk management.

Intrinsic bioremediation will be the preferred alternative when the costs of conventional remediation are high, the problem compounds are easily biodegradable, aquifer conditions are appropriate, there are no nearby groundwater receptors, and/or there is a well-defined surface-water discharge. Intrinsic bioremediation alone may not be the best alternative when the costs of conventional remediation are moderate to low and/or there is a large source

TABLE 10. Biodegradation results from field and laboratory studies at intrinsic bioremediation sites.

Site	Contaminant	V (m/d)	Field Results	Laboratory Results
Borden, Ontario (Barker et al. 1987)	BTX stock solution injected into uncontaminated aquifer	0.09	Zero-order decay rates from mass balance method: benz. = 30 mg d⁻¹; tol. = 37 mg d⁻¹; m-xyl. = 47 mg d⁻¹; p-xyl. = 55 mg d⁻¹; o-xyl. = 33 mg d⁻¹.	Zero-order decay rates (per 1,800 L) from aerobic microcosms: benz. = 58 d⁻¹; tol. = 61 mg d⁻¹; m-xyl. = 50 mg d⁻¹; p-xyl. = 65 mg d⁻¹; o-xyl. = 54 mg d⁻¹.
Borden, Ontario (Barbaro et al. 1992)	stock solution contacted with gasoline then injected into leachate plume	0.09	% loss over 4 m travel.	
Rocky Point, NC (Borden et al. 1995)	residual gasoline from UST	0.08	Rates from conc. vs. travel time: benz. = 0.0002 d⁻¹; tol. = 0.0021 d⁻¹; e-benz. = 0.0015 d⁻¹; m,p-xyl. = 0.0013 d⁻¹; o-xyl. = 0.0021 d⁻¹.	Rates from Fe/SO₄ reducing microcosms: benz. = 0.024 d⁻¹; tol. = 0.045 d⁻¹; e-benz. = 0.002 d⁻¹; m,p-xyl. = 0.02 d⁻¹; o-xyl. = 0.056 d⁻¹.
Kalkaska, MI (Chiang et al. 1989)	natural gas condensate–BTEX	0.2	Rates from mass balance: benz. = 0.0095 d⁻¹.	Rates from aerobic microcosms: BTX = 0.01 to 0.1 d⁻¹.
Columbus, MS (MacIntyre et al. 1993)	stock solution of benzene, p-xylene, naphthalene, o-dichlorobenzene		Tritium used as nonreactive tracer. Mineralization proven using ¹⁴C -p-xyl. Rates from mass balance: benz. = 0.0070 d⁻¹; p-xyl. = 0.0107 d⁻¹; naphthalene = 0.0064 d⁻¹; o-DCB = 0.0046 d⁻¹.	
Sleeping Bear, MI (Wilson et al. 1994b; Schafer, 1994)	residual gasoline from UST release–BTEX	0–0.4	Rates from conc. vs. travel time using 2,3-dimethylpentane as an internal standard: benz. = N.S.; tol. = 0.02 - 0.07 d⁻¹; e-benz. = 0.03 - 0.011 d⁻¹; m-xyl. = 0.004 - 0.014 d⁻¹; p-xyl. = 0.002 - 0.010 d⁻¹; o-xyl. = 0.004 - 0.011 d⁻¹.	Rates from methanogenic microcosms: benz. = N.S.; tol. = 0.007 - 0.04 d⁻¹; e-benz. = N.S.; mp-xyl. = N.S.; o-xyl. = N.S. 12 to 16 mg/L CH₄ produced in lab microcosms.
Indian River, FL (Kemblowski et al. 1987)	gasoline from UST–BTEX	0.06	Conc. vs. travel time: benz. = 0.0085 d⁻¹.	1st-order rates from aerobic microcosms: benz. = 0.02 to 0.2 d⁻¹.

TABLE 10. (continued).

Site	Contaminant	V (m/d)	Field Results	Laboratory Results
Morgan Hill, CA (Kemblowski et al. 1987)	gasoline–BTEX	0.05	Rates from conc. vs. travel time: benz. = 0.0035 d^{-1}.	
Eglin AFB, FL (Wilson et al. 1994a)	JP-4 from POL depot	1.3	Rates from conc. vs. travel time using 1,2,4-trimethylbenzene as internal standard: benz.= B.D.; tol. = 0.05 to 0.013 d^{-1}; e-benz. = 0.03 to 0.05 d^{-1}; m-xyl. = 0.02 to 0.1 d^{-1}; p-xyl. = 0.02 to 0.08 d^{-1}; o-xyl. = 0.21 d^{-1}.	
Hill AFB, UT (Wiedemeier et al. 1994)	JP-4 from POL depot	0.5	Rates from conc. vs. travel time using total trimethyl-benzene as internal standard: benz. = 0.03 to 0.09 d^{-1}; e-benz. = 0.01 to 0.08 d^{-1}; p-xyl. = 0.01 to 0.03 d^{-1}; m-xyl. = 0 to 0.03 d^{-1}; o-xyl. = 0 to 0.02 d^{-1}. Toluene rate not calculable.	
Patrick AFB, FL (Wiedemeier et al. 1994)	700 gal (2,650 L) unleaded gasoline from UST	0.13	Rates from conc. vs. travel time using total methane as internal standard: benz. = 0 to 0.004 d^{-1}; tol. = 0.0006 to 0.004 d^{-1}; e-benz. = 0.0001 to 0.004 d^{-1}; p-xyl. = 0.001 to 0.003 d^{-1}; m-xyl. = 0.001 to 0.004 d^{-1}; o-xyl. = 0.004 to 0.02 d^{-1}.	
Fairfax, VA (Buscheck et al. 1993)		0.015	Rates from conc. vs. travel time: benz. = 0.00055 d^{-1}; tol. = 0.00045 d^{-1}; e-benz. = 0.00045 d^{-1}; m,p,o-xyl. = 0.00040 d^{-1}.	
San Francisco, CA (Buscheck et al. 1993)		0.03	Rates from conc. vs. travel time: benz. = 0.0028 d^{-1}; tol. = 0.0022 d^{-1}; e-benz. = 0.0033 d^{-1}; m,p,o-xyl. = 0.0023 d^{-1}.	

TABLE 10. (continued).

Site	Contaminant	V (m/d)	Field Results	Laboratory Results
Alameda County, CA (Buscheck et al. 1993)	gasoline–BTEX	0.01	Rates from conc. vs. travel time: benz. = 0.0020 d^{-1}; tol. = 0.0017 d^{-1}; e-benz. = 0.0020 d^{-1}; m,p,o-xyl. = 0.0017 d^{-1}.	
Elko County, NV (Buscheck et al. 1993)	gasoline–BTEX	0.04	Rates from conc. vs. travel time: benz. = 0.001 d^{-1}.	
Sampson Co., NC (Borden, personal communication 1995)	gasoline from UST– BTEX/MTBE	0.04	High nitrate concentrations in groundwater may enhance biodegradation 1st-order rates from mass flux: MTBE = 0.0006 d^{-1}; benz. = 0.0006 d^{-1}; tol. = 0.0021 d^{-1}; e-benz. = 0.0023 d^{-1}; m,p-xyl. = 0.0016 d^{-1}; o-xyl. = 0.0009 d^{-1}.	Toluene and ethylbenzene rapidly degraded in denitrifying microcosms after a 56-d lag period.
Traverse City (Wilson et al. 1990)	aviation gasoline from UST– BTEX	1.5	Rates from conc. vs. travel time: benz. = 0.001 d^{-1}; tol. = 0.2 d^{-1}; m,p,o-xyl. = 0.004 d^{-1}.	Anaerobic microcosm rates: benz. = 0.07 d^{-1}; tol. = 0.04 d^{-1}; m,p-xyl. = 0.06 d^{-1}; o-xyl. = 0.07 d^{-1}. Methane produced in microcosms.
Broward Co., FL (Caldwell et al. 1992)	gasoline from UST–BTEX and MTBE	0.1	Anaerobic decay rate from matching BIOPLUME for total BTEX = 0.00012 d^{-1}. Aerobic decay will increase net biodegradation.	
Pensacola, FL (Bekins et al. 1993)	creosote–phenols	0.3 to 1.2	Selected phenols were completely degraded over a 100-d travel time through methanogenic aquifer.	Selected phenols were completely degraded over 100 to 200 d in methanogenic microcosms.
Bemidji, MN (Baedecker et al. 1993)	crude oil–BTEX	0.25	tol. and o-xyl. depleted over 20 m (200 d travel time); benz. and e-benz. depleted over 100 m. Downgradient migration was limited by mixing with uncontaminated water.	98% benz. loss in 125 d and 99% tol. loss in 45 d in anaerobic microcosms.

Intrinsic Bioremediation

TABLE 10. (continued).

Site	Contaminant	V (m/d)	Field Results	Laboratory Results
Perth, Australia (Thierrin et al. 1993)	gasoline from UST–BTEX	0.4	Rates from conc. vs. travel time: benz. = N.S.; tol. = 0.006 d⁻¹; e-benz. = 0.003 d⁻¹; m,p-xyl. = 0.004 d⁻¹; o-xyl. = 0.006 d⁻¹; naphthalene = 0.004 d⁻¹. Field (plume scale) rates closely match rates from tracer test using deuterated compounds.	Anaerobic columns with 14 mg/L SO₄, benz. = N.S. tol. = 2.3 d⁻¹ e-benz. = N.S. o-xyl. = N.S.
Manufacturing Plant (Davis et al. 1994)	benzene only	0.16	BIO1D match to field data showed benzene decay rate > 0.01 d⁻¹.	Over 90% benzene loss over 77 d in methanogenic and sulfate-reducing microcosms.
Cliffs-Dow (Klecka et al. 1990)	charcoal wastes, phenols, naphthalene	0.2 to 0.46	All organics degraded within 100 m of source.	All organics degraded in aerobic microcosms within 30 to 60 d.
Hill AFB-2, UT (Dupont et al. 1994)	18,000 gal (68,137 L) UST	0.14	Rates from mass balance: 1st order for TPH = 0.005 d⁻¹. Zero order for: benz. = 0.02 kg/d; e-benz. = 0.06 kg/d; p-xyl. = 0.06 kg/d.	
Gas Plant (Piontek et al. 1994)	NAPL released from natural gas plant–BTEX		105 reduction in BTEX over 100 m.	
Picatinny Arsenal, NJ (Martin and Imbrigiotta 1994)	TCE, 1,1,1-trichloroethane and metals from plating wastewater	0.3 to 1.0	Spatial distribution of TCE, DCE, and VC indicate reductive dechlorination.	Anaerobic microcosms: TCE = 0.0001 to 0.003 d⁻¹.
St. Joseph, MI (Wilson et al. 1994b)	TCE from lagoons/dry wells	0.1	Rates from mass flux: TCE = 0.001 to 0.003 d⁻¹.	
Finger Lakes, NY (Major et al. 1994)	TCE, acetone, methanol		Spatial distribution of TCE, DCE, VC, and ethene were indicative of reductive dechlorination.	

(a) benz. = benzene, tol. = toluene, e-benz. = ethylbenzene, xyl. = xylenes, TPH = total petroleum hydrocarbons, N.S. = not significant, POL = petroleum, oil, and lubricants.

of poorly degradable compounds. Norris et al. (1994) argue that, at some sites, it may be more cost effective to implement some low-level activities (e.g., limited air sparging or venting) than to rely on intrinsic bioremediation as the only management technique. If these limited remediation activities can shorten the monitoring period by several years, they may more than pay for themselves in reduced long-term monitoring costs.

ACKNOWLEDGMENTS

Portions of the research described in this article were supported by the American Petroleum Institute under Grant No. GW-25A-0400-38. The opinions, findings, and conclusion expressed are those of the authors and do not necessarily represent those of the American Petroleum Institute.

REFERENCES

Acton, D. W., and J. F. Barker. 1992. "In Situ Biodegradation Potential of Aromatic Hydrocarbons in Anaerobic Groundwaters." *Journal of Contaminant Hydrology 9*: 325-352.

Baedecker, M. J., I. M. Cozzarelli, R. P. Eaganhouse, D. I. Siegel, and P. C. Bennett. 1993. "Crude Oil in a Shallow Sand and Gravel Aquifer: III. Biogeochemical Reactions and Mass Balance Modeling in Anoxic Groundwater." *Applied Geochemistry 8*: 569-586.

Barbaro, J. R., J. F. Barker, L. A. Lemon, and C. I. Mayfield. 1992. "Biotransformation of BTEX Under Anaerobic Denitrifying Conditions: Field and Laboratory Observations." *Journal of Contaminant Hydrology 11*: 245-272.

Barker, J. F., G. C. Patrick, and D. Major. 1987. "Natural Attenuation of Aromatic Hydrocarbons in a Shallow Sand Aquifer." *Water Monitoring Review, Winter*: 64-71.

Barker, J. F., and J. T. Wilson. 1992. "Natural Biological Attenuation of Aromatic Hydrocarbons Under Anaerobic Conditions." In *Proceedings of the EPA Subsurface Restoration Conference*, Dallas, TX, pp. 57-58.

Beck, M. M. 1994. "Natural Attenuation of Chlorinated Solvents at Dover Air Force Base." In *Proceedings of the U. S. Air Force Center of Excellence Third Environmental Restoration Technology Transfer Symposium*, Nov. 8 - 10, 1994, San Antonio, TX.

Bedient, P. B., H. S. Rifai, and C. J. Newell. 1994. *Ground Water Contamination: Transport and Remediation.* Prentice Hall, Engelwood Cliffs, NJ.

Bekins, B. A., E. M. Godsy, and D. F. Goerlitz. 1993. "Modeling Steady-State Methanogenic Degradation of Phenols in Groundwater." *Journal of Contaminant Hydrology 14*: 279-294.

Borden, R. C., and P. B. Bedient. 1986. "Transport of Dissolved Hydrocarbons Influenced by Oxygen-Limited Biodegradation: 1. Theoretical Development." *Water Resources Res. 13*: 1973-1982.

Borden, R. C. 1990. "Subsurface Monitoring Data for Assessing In-Situ Biodegradation of Aromatic Hydrocarbons (BTEX) in Groundwater." Proposal to the American Petroleum Institute.

Borden, R. C., C. A. Gomez, and M. T. Becker. 1994. "Natural Bioremediation of a Gasoline Spill." In R. E. Hinchee et al. (Eds.), *Hydrocarbon Bioremediation*, pp. 290-295. Lewis Publishers, Boca Raton, FL.

Borden, R. C., C. A. Gomez, and M. T. Becker. 1995. "Geochemical Indicators of Intrinsic Bioremediation." Submitted to *Ground Water.*

Buscheck, T. E., K. T. O'Reilly, and S. N. Nelson. 1993. "Evaluation of Intrinsic Bioremediation at Field Sites." In *Proceedings of the 1993 Petroleum Hydrocarbons and Organic Chemicals in Ground Water: Prevention, Detection, and Restoration*, pp. 367-381. Water Well Journal Publishing Co., Dublin, OH.

Caldwell, K. R., D. L. Tarbox, K. D. Barr, S. Fiorenza, L. E. Dunlap, and S. B. Thomas. 1992. "Assessment of Natural Bioremediation as an Alternative to Traditional Active Remediation at Selected Amoco Oil Company Sites, Florida." In *Proceedings of the 1992 Petroleum Hydrocarbons and Organic Chemicals in Ground Water: Prevention, Detection, and Restoration*, pp. 509-525. Well Water Journal Publishing Co., Dublin, OH.

Celia, M. A., J. S. Kindred, and I. Herrera. 1989. "Contaminant Transport and Biodegradation: 1. A Numerical Model for Reactive Transport in Porous Media." *Water Resources Res.* 25(6): 1141- 148.

Chiang, C. Y., J. P. Salanitro, E. Y. Chai, J. D. Colthart, and C. L. Klein. 1989. "Aerobic Biodegradation of Benzene, Toluene, and Xylene in a Sandy Aquifer: Data Analysis and Computer Modeling." *Ground Water* 27: 823-834.

Connor, J. A., C. J. Newell, P. Nevin, and H. S. Rifai. 1994. "Guidelines for Use of Groundwater Spreadsheet Models in Risk-Based Corrective Action Design." In *Proceedings of the 1994 Petroleum Hydrocarbons and Organic Chemicals in Ground Water: Prevention, Detection, and Remediation*, pp. 43-51. Ground Water Publishing Co., Dublin, OH.

Cozzarelli, I. M., and M. J. Baedecker. 1992. "Oxidation of hydrocarbons coupled to reduction of inorganic species in groundwater." In Y. K. Kharaka and A. S. Maest (Eds.), *Water-Rock Interaction*, pp. 275-278. A. A. Balkema, Rotterdam.

Cozzarelli, I. M., R. P. Eaganhouse, and M. J. Baedecker. 1990. "Transformation of Monoaromatic Hydrocarbons to Organic Acids in Anoxic Ground-Water Environment." *Environ. Geol. Water Sci.* 16: 135-141.

Davis, J. W., N. J. Klier, and C. L. Carpenter. 1994. "Natural Biological Attenuation of Benzene in Ground Water Beneath a Manufacturing Facility." *Ground Water* 32(2): 215-226.

de Bruin, W. P., M. J. J. Kotterman, M. A. Posthumus, G. Schraa, and A. J. B. Zehnder. 1992. "Complete Biological Reductive Transformation of Tetrachloroethene to Ethane." *Applied Environmental Microbiology* 58(6): 1996-2000.

Domenico, P. A. 1987. "An Analytical Model for Multidimensional Transport of a Decaying Contaminant Species." *Journal of Hydrology* 91: 49-59

Dupont, R. R., D. L. Sorensen, and M. Kemblowski. 1994. "Evaluation of Intrinsic Bioremediation at an Underground Storage Tank Site in Northern Utah." In *Proceedings of the EPA Symposium on Intrinsic Bioremediation of Ground Water*, pp. 176-177. U.S. Environmental Protection Agency, Washington, DC.

Edwards, E. A., L. E. Wills, D. Grbic-Galic, and M. Reinhard. 1991. "Anaerobic Degradation of Toluene and Xylene—Evidence for Sulphate as the Terminal Electron Acceptor." In R. E. Hinchee and R. F. Olfenbuttel (Eds.) *In Situ Bioreclamation: Applications and Investigations for Hydrocarbon and Contaminated Site Remediation*, pp. 463-471. Butterworth-Heinemann, Stoneham, MA.

Evans, P. J., D. T. Mang, and L. Y. Young. 1991. "Degradation of Toluene and *m*-Xylene and Transformation of *o*-Xylene by Denitrifying Enrichment Cultures." *Applied Environmental Microbiology* 57: 450-454.

Fiorenza, S., E. L. Hockman, Jr., S. Szojka, R. M. Woeller, and J. W. Wigger. 1994. "Natural Anaerobic Degradation of Chlorinated Solvents at a Canadian Manufacturing Plant." In R. E. Hinchee, A. Leeson, L. Semprini, and S. K. Ong (Eds.), *Bioremediation of Chlorinated and Polycyclic Aromatic Hydrocarbon Compounds*, pp. 277-286. Lewis Publishers, Inc., Boca Raton, FL.

Freedman, D. L., and J. M. Gossett. 1989. "Biological Reductive Transformation of Tetrachloroethylene and Trichlorethylene to Ethylene Under Methanogenic Conditions." *Applied Environmental Microbiology* 55(9): 2144-2151.

Godsy, E. M., D. F. Goerlitz, and D. Grbic-Galic. 1992. "Methanogenic Biodegradation of Creosote Contaminants in Natural and Simulated Ground-Water Ecosystems." *Ground Water 30*(2): 232-242.

Grbic-Galic, D., and T. M. Vogel. 1987. "Transformation of Toluene and Benzene by Mixed Methanogenic Cultures." *Applied Environmental Microbiology 53*: 254-260.

Haston, Z. C., P. K. Sharma, J. N. Black, and P. L. McCarty. 1994. "Enhanced Reductive Dechlorination of Chlorinated Ethenes." Presented at *Symposium on Bioremediation of Hazardous Wastes,* San Francisco, CA. U.S. Environmental Protection Agency.

Hunt, M. J., M. A. Beckman, M. A. Barlaz, and R. C. Borden. 1995. "Anaerobic BTEX Biodegradation in Laboratory Microcosms and In-Situ Columns." Submitted to *Proceedings of the 1995 Petroleum Hydrocarbons and Organic Chemicals in Ground Water: Prevention, Detection, and Restoration.* Well Water Journal Publishing Co., Dublin, OH.

Kemblowski, M. W., J. P. Salanitro, G. M. Deeley, and C. C. Stanley. 1987. " Fate and Transport of Residual Hydrocarbon in Groundwater: A Case Study." In *Proceedings of Petroleum Hydrocarbons and Organic Chemicals in Ground Water: Prevention, Detection, and Restoration: A Conference and Exposition,* pp. 207-231. National Water Well Association, Dublin, OH.

Kinzelbach, W., W. Schafer, and J. Herzer. 1991. "Numerical Modeling of Natural and Enhanced Denitrification Processes in Aquifers." *Water Resources Res.* 27(6): 1123-1135.

Kitanidis, P. K., L. Semprini, D. H. Kampbell, and J. T. Wilson. 1993. "Natural Anaerobic Bioremediation of TCE at the St. Joseph, Michigan Superfund Site." *Proceedings of the EPA Symposium on Bioremediation of Hazardous Wastes: Research, Development, and Field Evaluations,* pp. 47-50. U.S. Environmental Protection Agency, EPA/600/R-93/054.

Klecka, G. M., J. W. Davis, D. R. Gray, and S. S. Madsen. 1990. "Natural Bioremediation of Organic Contaminants in Ground Water: Cliffs-Dow Superfund Site." *Ground Water 28*(4): 534-543.

Kuhn, E. P., and J. M. Suflita. 1989a. "Dehalogenation of Pesticides by Anaerobic Microorganisms in Soils and Ground Water: A Review." In B. L. Sawhney and K. Brown (Eds.), *Reactions and Movement of Organic Chemicals in Soils,* Special Publication 22, pp. 111-180. Soil Science Society of America, Madison, WI.

Kuhn, E. P., and J. Suflita. 1989b. "Anaerobic Biodegradation of Nitrogen-substituted and Sulfonated Benzene Aquifer Contaminants." *Hazardous Waste Hazardous Materials 6*(2): 121-133.

Lovley, D. R., and D. J. Lonergan. 1990. "Anaerobic Oxidation of Toluene, Phenol, and *p*-Cresol by the Dissimilatory Iron-Reducing Organism, GS-15." *Applied Environmental Microbiology 56*(1): 858-864.

MacIntyre, W. G., M. Boggs, C. P. Antworth, and T. B. Stauffer. 1993. "Degradation Kinetics of Aromatic Organic Solutes Introduced Into a Heterogeneous Aquifer." *Water Resources 29*(12): 4045-51.

MacQuarrie, K. T. B., E. A. Sudicky, and E. O. Frind. 1990. "Simulation of Biodegradable Organic Contaminants in Groundwater: 1. Numerical Formulation in Principal Directions." *Water Resources Res.* 26(2): 207-222.

Major, D., E. Cox, E. Edwards, and P. W. Hare. 1994. "The Complete Dechlorination of Trichloroethene to Ethene Under Natural Conditions in a Shallow Bedrock Aquifer Located in New York State." In *Proceedings of the EPA Symposium on Intrinsic Bioremediation of Ground Water,* pp. 187-189. U.S. Environmental Protection Agency, Washington, DC.

Major, D. W., E. W. Hodgins, and B. J. Butler. 1991. "Field and Laboratory Evidence of In Situ Biotransformation of Tetrachloroethene to Ethene and Ethane at a Chemical Transfer Facility in North Toronto." In R. E. Hinchee and R. F. Olfenbuttel (Eds.), *On-Site Bioreclamation,* pp. 147-171. Butterworth-Heinemann, Stoneham, MA.

Major, D. W., C. I. Mayfield, and J. F. Barker. 1988. "Biotransformation of Benzene by Denitrification in Aquifer Sand." *Ground Water 26*: 8-14.

Martin, M. and T.E. Imbrigiotta. 1994. "Contamination of Ground Water With Tri-chloroethylene at the Building 24 Site at Picatinny Arsenal, New Jersey." In *Proceedings of the EPA Symposium on Intrinsic Bioremediation of Ground Water*, pp. 143-153. U.S. Environmental Protection Agency, Washington, DC.

McCarty, P. L. 1994. "An Overview of Anaerobic Transformation of Chlorinated Solvents." *Proceedings of the EPA Symposium on Intrinsic Bioremediation of Ground Water*. U.S. Environmental Protection Agency, EPA/540/R-94/515, August 30-September 1, Denver, CO.

McCarty, P. L., L. Semprini. M. E. Dolan, T. C. Harmon, C. Tiedeman, S. M. Gorelick. 1991. "In Situ Methanotrophic Bioremediation for Contaminated Water at St. Joseph, Michigan." In R. E. Hinchee and R. G. Olfenbuttel (Eds.), *On-Site Bioreclamation Processes for Xenobiotic and Hydrocarbon Treatment*, pp. 16-40. Butterworth-Heinemann, Stoneham, MA.

McCarty, P. L. and J. T. Wilson. 1992. "Natural Anaerobic Treatment of a TCE Plume, St. Joseph, Michigan, NPL Site." *Symposium on Bioremediation of Hazardous Wastes*, pp. 47-50. U.S. Environmental Protection Agency EPA/600/R-92/126.

Molz, F. J., M. A. Widdowson, and L. D. Benefield. 1986. "Simulation of Microbial Growth Dynamics Coupled to Nutrient and Oxygen Transport in Porous Media." *Water Resources Res.* 22(8): 1207-1216.

Norris, R. D., J. C. Dey, and D. P. Shine. 1994. "The Use of Low Level Activities To Assist Intrinsic Bioremediation." In *Proceedings of the EPA Symposium on Intrinsic Bioremediation of Ground Water*, pp. 173-174. U.S. Environmental Protection Agency, Washington, DC.

Odencrantz, J. E., A. J. Valocchi, and B. E. Rittman. 1990. "Modeling Two-dimensional Solute Transport with Different Biodegradation Kinetics." In *Proceedings of Petroleum Hydrocarbons and Organic Chemicals in Ground Water: Prevention, Detection and Restoration.* National Water Well Association, Houston, TX.

Piontek, K., T. Sale, S. de Albuquerque, and J. Cruze. 1994. "Demonstrating Intrinsic Bioremediation of BTEX at a Natural Gas Plant." In *Proceedings of the EPA Symposium on Intrinsic Bioremediation of Ground Water*, pp. 179-180. U.S. Environmental Protection Agency, Washington, D.C.

Rifai, H. S., P. B. Bedient, J. T. Wilson, K. M. Miller, and J. M. Armstrong. 1988. "Biodegradation Modeling at an Aviation Fuel Spill." *ASCE Journal of Environmental Engineering* 114(5): 1007-1029.

Rifai, H. S. 1990. "Assessing In-Situ Biodegradation of BTEX in Groundwater." Project workplan, American Petroleum Institute, unpublished.

Rifai, H. S. 1994. "Mathematical Modeling of Intrinsic Bioremediation at Field Sites." In *Proceedings of the EPA Symposium on Intrinsic Bioremediation of Ground Water*. U.S. Environmental Protection Agency, EPA/540/R-94/515, August 30-September 1, Denver, CO.

Rifai, H. S., and L. P. Hopkins. 1995. "Natural Attenuation Toolbox." Submitted to American Petroleum Institute.

Rifai, H. S., C. J. Newell, R. Miller, S. Taffinder, and M. Rounsaville. 1995. "Simulation of Natural Attenuation with Multiple Electron Acceptors." In R.E. Hinchee, J. T. Wilson, and D. C. Downey (Eds.), *Intrinsic Bioremediation*, pp. 53-58. Battelle Press, Columbus, OH. (In Press)

Schafer, Michael Blaine. 1994. "Methanogenic Biodegradation of Alkylbenzenes in Groundwater From Sleeping Bear Dunes National Lakeshore." M.S. Thesis, North Carolina State University, Raleigh, NC.

Semprini, L., and P. L. McCarty. 1991. "Comparison between Model Simulations and Field Results for in Situ Biorestoration of Chlorinated Aliphatics: Part 1. Biostimulation of Methanotrophic Bacteria." *Ground Water* 29(3): 365-374.

Srinivasan, P., and J. W. Mercer. 1988. "Simulation of Biodegradation and Sorption Processes in Ground Water." *Ground Water* 26(4): 475-487.

Thierrin, J., G. B. Davis, C. Barber, B. M. Patterson, F. Pribac, T. R. Power, and M. Lambert. 1993. "Natural degradation rates of BTEX compounds and napthalene in a sulfate reducing groundwater environment." *Hydrological Sciences Journal* 38(4): 309-322.

Tucker, W. A., C. T. Huang, J. M. Bral, and R. E. Dickinson. 1986. "Development and Validation of the Underground Leak Transport Assessment Model (ULTRA)." In *Proceedings of Petroleum Hydrocarbons and Organic Chemicals in Ground Water: Prevention, Detection and Restoration*, pp. 53-75. National Well Association, Houston, TX.

Vogel, T. M., and D. Grbic-Galic. 1986. "Incorporation of Oxygen from Water Into Toluene and Benzene During Anaerobic Fermentative Transformation." *Applied Environmental Microbiology* 52: 200-202.

Wheeler, M. F., C. N. Dawson, P. B. Bedient, C. Y. Chiang, R. C. Borden, and H. S. Rifai. 1987. "Numerical Simulation of Microbial Biodegradation of Hydrocarbons in Ground Water." In *Proceedings of the Solving Ground Water Problems with Models Conference*. National Water Well Association, Dublin, OH.

Wiedemeier, T. H., D. C. Downey, J. T. Wilson, D. H. Kampbell, R. N. Miller and J. E. Hansen. 1994. "Technical Protocol for Implementing the Intrinsic Remediation With Long Term Monitoring Option for Natural Attenuation of Dissolved Phase Fuel Contamination in Ground Water." Air Force Center for Environmental Excellence Publication (draft).

Wiedemeier, T. H., M. A. Swanson, J. T. Wilson, D. H. Kampbell, R. N. Miller, and J. E. Hansen. 1995a. "Patterns of Intrinsic Bioremediation at Two United States Air Force Bases." Submitted to *Proceedings of the 1995 Petroleum Hydrocarbons and Organic Chemicals in Ground Water: Prevention, Detection, and Restoration*. Well Water Journal Publishing Co., Dublin, OH.

Wiedemeier, T. H., M. A. Swanson, J. T. Wilson, D. H. Kampbell, R. N. Miller, and J. E. Hansen. 1995b. "Patterns of Intrinsic Bioremediation at Two U.S. Air Force Bases." In R. E. Hinchee, J. T. Wilson, and D. C. Downey (Eds.), *Intrinsic Bioremediation*, pp. 31-51. Battelle Press, Columbus, OH.

Wilson, J. T., F. M. Pfeffer, J. W. Weaver, D. H. Kampbell, T. H. Wiedemeier, J. E. Hansen, and R. N. Miller. 1994a. "Intrinsic Bioremediation of JP-4 Jet Fuel." In *Proceedings of the EPA Symposium on Intrinsic Bioremediation of Ground Water*, p. 189. U.S. Environmental Protection Agency, EPA/540/R-94/515.

Wilson, B. H., G. B. Smith, and J. F. Rees. 1986. "Biotransformations of Selected Alkylbenzenes and Halogenated Aliphatic Hydrocarbons in Methanogenic Aquifer Material: A Microcosms Study." *Environmental Science Technology* 20: 997-1002.

Wilson, J. T., J. W. Weaver, and D. H. Kampbell. 1994b. "Intrinsic Bioremediation of TCE in Ground Water at an NPL site in St. Joseph, Michigan." In *Proceedings of the EPA Symposium on Intrinsic Bioremediation of Ground Water*, August 30-September 1, Denver, CO. U.S. Environmental Protection Agency, EPA/540/R-94/515.

Wilson, B. H., J. T. Wilson, D. H. Kampbell, B. E. Bledsoe, and J. M. Armstrong. 1990. "Biotransformation of Monoaromatic and Chlorinated Hydrocarbons at an Aviation Gasoline Spill Site." *Geomicrobiology Journal* 8: 225-240.

Patterns of Intrinsic Bioremediation at Two U.S. Air Force Bases

Todd H. Wiedemeier, Matthew A. Swanson, John T. Wilson, Donald H. Kampbell, Ross N. Miller, and Jerry E. Hansen

ABSTRACT

Intrinsic bioremediation of benzene, toluene, ethylbenzene, and xylenes (BTEX) occurs when indigenous microorganisms work to reduce the total mass of contamination in the subsurface without the addition of nutrients. A conservative tracer, such as trimethylbenzene, found commingled with the contaminant plume can be used to distinguish between attenuation caused by dispersion, dilution from recharge, volatilization, and sorption and attenuation caused by biodegradation. Patterns of intrinsic bioremediation can vary markedly from site to site depending on governing physical, biological, and chemical processes. Intrinsic bioremediation causes measurable changes in groundwater chemistry. Specifically, concentrations of contaminants, dissolved oxygen, nitrate, ferrous iron, sulfate, and methane in groundwater change both temporally and spatially as biodegradation proceeds Operations at Hill Air Force Base (AFB) and Patrick AFB resulted in fuel-hydrocarbon contamination of soil and groundwater. In both cases, trimethylbenzene data confirm that dissolved BTEX is biodegrading. Geochemical evidence from the Hill AFB site suggests that aerobic respiration, denitrification, iron reduction, sulfate reduction, and methanogenesis all are contributing to intrinsic bioremediation of dissolved BTEX. Sulfate reduction is the dominant biodegradation mechanism at this site. Geochemical evidence from Patrick AFB suggests that aerobic respiration, iron reduction, and methanogenesis are contributing to intrinsic bioremediation of dissolved BTEX. Methanogenesis is the dominant biodegradation mechanism at this site.

INTRODUCTION

Microorganisms obtain energy for cell production and maintenance by facilitating the transfer of electrons from electron donors to electron acceptors. This results in the oxidation of the electron donor and the reduction of the electron acceptor. Common electron donors at fuel hydrocarbon-contaminated sites are natural organic carbon and fuel-related organic compounds, including BTEX. Electron acceptors are elements or compounds that occur in relatively oxidized form. Common electron acceptors found in groundwater include dissolved oxygen, nitrate, ferric iron (present as ferric orthohydroxide), sulfate, and carbon dioxide. Microorganisms generally utilize electron acceptors in a preferred order while metabolizing fuel hydrocarbons (Bouwer 1992). Depending on the types of electron acceptors and nutrients present, pH conditions, and alkalinity, biodegradation can occur via aerobic respiration, denitrification, ferric iron reduction, sulfate reduction, or methanogenesis. Dissolved oxygen is utilized as the first and primary electron acceptor. After the dissolved oxygen is consumed, anaerobic microorganisms typically use electron acceptors in the following order of preference: nitrate, ferric iron, sulfate, and finally carbon dioxide. Environmental conditions and microbial competition will ultimately determine which processes will dominate at a given site. Vroblesky and Chapelle (1994) show that the dominant terminal electron accepting process can vary both temporally and spatially in an aquifer contaminated with fuel hydrocarbons.

Three lines of evidence can be used to document the occurrence of natural attenuation and intrinsic bioremediation (National Research Council 1993): (1) documented loss of contaminants at the field scale, (2) geochemical evidence, and (3) microcosm studies. The first line of evidence, documented loss of contaminants at the field scale, requires historical data. The second line of evidence, groundwater geochemistry, can be used to determine the relative importance of each of the mechanisms of natural attenuation and the mechanisms of intrinsic bioremediation that are operating at a site. The third line of evidence, the microcosm study, can be used to estimate rates of biodegradation. Other methods for determining biodegradation rates rely on chemical evidence and include use of a conservative tracer (Wiedemeier et al. 1995), or interpretation of a steady-state contaminant plume configuration (Buscheck and Alcantar 1995).

The simplified stoichiometry of benzene biodegradation involving common electron acceptors is shown in Table 1. The stoichiometry presented in Table 1 assumes that no cellular mass production occurs and therefore could be conservative by a factor of up to three. The stoichiometry of toluene, ethylbenzene, and xylene biodegradation is similar to that for benzene. These reactions cause measurable changes in groundwater chemistry. Evidence of these changes can be used to support the occurrence of intrinsic bioremediation and to determine which mechanisms of intrinsic bioremediation are most important at a given site.

PATTERNS OF INTRINSIC BIOREMEDIATION AT HILL AIR FORCE BASE, UTAH

Previous investigations at the petroleum, oil, and lubricants (POL) facility at Hill AFB, Utah, determined that JP-4 jet fuel had been released into the soil

TABLE 1. Stoichiometry of common biodegradation reactions.

Benzene Biodegradation Reactions	Mass Ratio of Electron Acceptor to Benzene	Mass Ratio of Metabolic By-Product to Benzene	Average Mass Ratio of Electron Acceptor to Total BTEX [a]	Average Mass Ratio of Metabolic By-Product to Total BTEX [a]	Mass of BTEX Degraded per Unit Mass of Electron Acceptor Utilized (mg) [a]	Mass of BTEX Degraded per Unit Mass of Metabolic By-Product Produced (mg) [a]
$7.5O_2 + C_6H_6 \Rightarrow 6CO_{2,g} + 3H_2O$ Benzene oxidation /aerobic respiration	3.1:1	—	3.14:1	—	0.32	—
$6NO_3^- + 6H^+ + C_6H_6 \Rightarrow 6CO_{2,g} + 6H_2O + 3N_{2,g}$ Benzene oxidation / denitrification	4.8:1	—	4.9:1	—	0.21	—
$60H^+ + 30Fe(OH)_3 + C_6H_6 \Rightarrow 6CO_2 + 30Fe^{2+} + 78H_2O$ Benzene oxidation / iron reduction	41.1:1	21.5:1	—	21.8:1	—	0.05
$7.5H^+ + 3.75SO_4^{2-} + C_6H_6 \Rightarrow 6CO_{2,g} + 3.75H_2S^\circ + 3H_2O$ Benzene oxidation / sulfate reduction	4.6:1	—	4.7:1	—	0.21	—
$4.5H_2O + C_6H_6 \Rightarrow 2.25CO_{2,g} + 3.75CH_4$ Benzene oxidation / methanogenesis	—	0.77:1	—	0.78:1	—	1.28

(a) Simple average of all BTEX compounds based on individual compound stoichiometry. This stoichiometry assumes no cellular mass is produced.

and shallow groundwater. The chronology of the JP-4 spill or spills is not known. The facility began operating in the early 1950s. Site characterization methods used to evaluate intrinsic bioremediation included Geoprobe® sampling of groundwater, cone penetrometer testing (CPT), soil borehole drilling, soil sample collection and analysis, monitoring well installation, sampling and analysis of groundwater from monitoring wells, and aquifer testing.

Site Geology and Hydrogeology

Hill AFB is located on a bench of the Wasatch Mountains on the edge of the Great Salt Lake Basin. Surface topography at the site slopes to the southwest. Shallow sediments consist of light reddish-brown to dark gray, cohesive clayey silts to silty clays. This unit is 1.2 to 4.6 m thick and is underlain by poorly to moderately sorted, yellowish-brown to reddish-brown, silty fine-grained sands that coarsen downward into a 0.9- to 6.7-m-thick sequence of moderately sorted, medium- to coarse-grained sands. Underlying the sands is a sequence of competent, thinly interbedded clay to silty clay and fine- to very fine-grained, clayey sand and silt of unknown thickness. This sequence of interbedded clay and fine-grained sand and silt acts as an effective barrier to the downward migration of water and contaminants, as indicated by geochemical data. Upward hydraulic gradients in the area also prevent downward migration of the contaminant plume.

The water table aquifer is present in the medium- to coarse-grained sands described above. The water table is present between 1.5 and 6.1 m below ground surface (bgs), and groundwater flow is to the southwest with an average horizontal gradient of 0.048 m/m. Available data suggest that there is almost no seasonal variation in groundwater flow direction or gradient at the site. Based on slug tests and pumping tests, the average hydraulic conductivity for the shallow medium- to coarse-grained sands of the shallow saturated zone is 0.0084 cm/s. With a gradient of 0.046 m/m, a hydraulic conductivity of 0.0084 cm/s, and an assumed effective porosity of 0.25, the average advective groundwater velocity is 1.34 m/day, or approximately 488 m/year. Because of the low total organic carbon (TOC) concentration and clay mineral content observed in the shallow saturated zone at this site, retardation of the BTEX compounds due to sorption is not expected to be a significant process affecting solute transport.

Figures 1a through 1f show the POL facility and the immediately adjacent area. These figures include data collected from 12 monitoring wells in the source area north of 6th Street in December 1993/January 1994. These wells cover a very small area relative to the extent of the dissolved plume, and these data represent the only data available for this area.

Nature and Extent of Contamination

Figure 1a shows the approximate extent of mobile light nonaqueous-phase liquid (LNAPL) at the site, the main source of BTEX dissolved in groundwater.

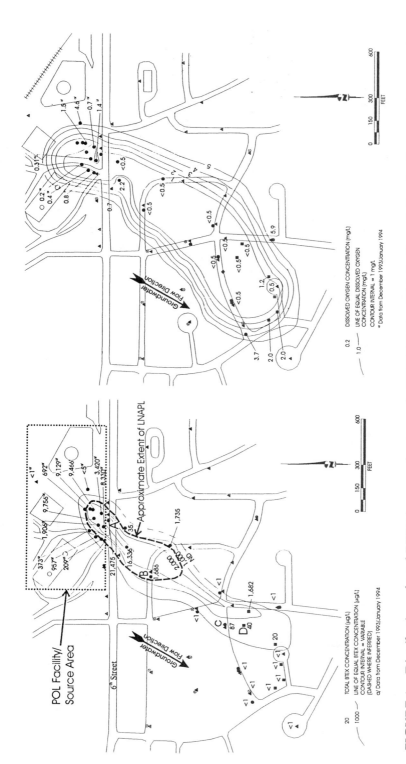

FIGURE 1a. Distribution of total BTEX in groundwater and approximate extent of LNAPL at Hill AFB, Utah (where 1 ft = 0.3 m).

FIGURE 1b. Distribution of dissolved oxygen in groundwater at Hill AFB, Utah (where 1 ft = 0.3 m).

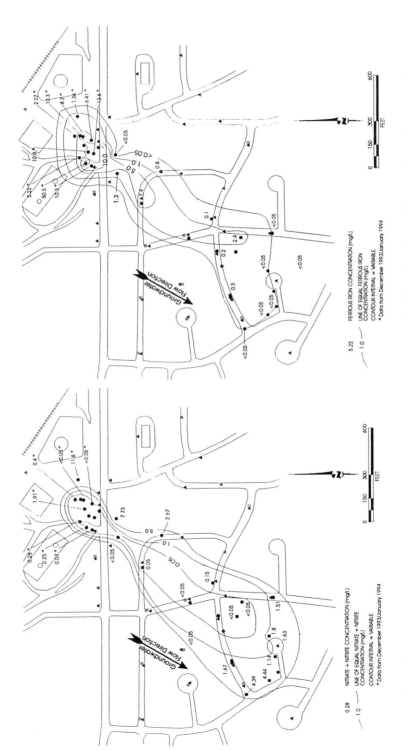

FIGURE 1c. Distribution of nitrate + nitrite in groundwater at Hill AFB, Utah (where 1 ft = 0.3 m).

FIGURE 1d. Distribution of ferrous iron in groundwater at Hill AFB, Utah (where 1 ft = 0.3 m).

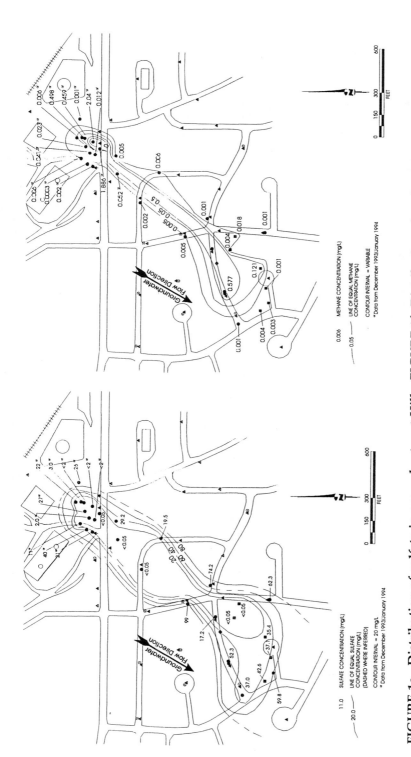

FIGURE 1e. Distribution of sulfate in groundwater at Hill AFB, Utah (where 1 ft = 0.3 m).

FIGURE 1f. Distribution of methane in groundwater at Hill AFB, Utah (where 1 ft = 0.3 m).

The LNAPL plume is composed of weathered JP-4 released from the POL facility. Measured residual total BTEX concentrations in soil decrease rapidly outside the area of mobile LNAPL contamination. The highest dissolved BTEX concentration observed in groundwater during this study was 21,475 µg/L. Figure 1a is also an isopleth map that shows the distribution of total BTEX dissolved in groundwater in July 1994. Dissolved BTEX contamination is migrating to the southwest, in the direction of groundwater flow.

Documented Loss of Contaminants at the Field Scale

Dissolved BTEX data collected in 1992, 1993, and 1994 indicate that the dissolved BTEX plume has reached steady-state conditions, and although fluctuations in contaminant concentrations were observed during this period, the plume is not expanding or migrating further downgradient. Based on the calculated advective velocity of the groundwater, the contaminant plume should have migrated approximately 442 m downgradient between August 1993 and July 1994. Available geochemical data suggest that the stabilization of the BTEX plume such a short distance downgradient of the source is primarily the result of intrinsic bioremediation, as discussed in the following sections.

To determine what portion of observed decreases in contaminant concentrations can be attributed to biodegradation, measured BTEX concentrations must be corrected for the effects of dispersion, dilution from recharge, volatilization, and sorption. A convenient way of doing this is to use a compound present in the dissolved hydrocarbon plume that has sorptive and volatilization properties similar to those of BTEX and that is recalcitrant under anaerobic conditions. One tracer compound that is useful in some, but not all, groundwater environments is trimethylbenzene (TMB). The three isomers of this compound (1,2,3-TMB, 1,2,4-TMB, and 1,3,5-TMB) have Henry's law constants and soil sorption coefficients that are similar to the BTEX compounds. Also, the TMB isomers are generally present in sufficient quantities in fuel mixtures to be readily detectable when dissolved in groundwater and are fairly recalcitrant in the anaerobic portion of the plume. The degree of recalcitrance of TMB is site-specific and the use of this compound as a tracer must be evaluated on a case-by-case basis.

The corrected concentration of a compound is the concentration of the compound that would be expected at one point (B) located downgradient of another point (A) after correcting for the effects of dispersion, dilution from recharge, volatilization, and sorption between points A and B. One method of calculating the corrected concentration is given by (Wiedemeier et al. 1994):

$$C_{B,corr} = C_B \left(\frac{TMB_A}{TMB_B} \right) \qquad (1)$$

Where $C_{B,corr}$ = corrected concentration of compound of interest at Point B
$\quad\ C_B$ = measured concentration of compound of interest at Point B

$$TMB_A \quad = \quad \text{measured concentration of trimethylbenzene at Point A}$$
$$TMB_B \quad = \quad \text{measured concentration of trimethylbenzene at Point B}$$

TMB is slightly more hydrophobic than the BTEX compounds and is not entirely recalcitrant under anaerobic conditions. However, if any TMB mass is lost to the processes of biodegradation or preferential sorption, the relationship shown above is more conservative (i.e., a lower mass lost due to biodegradation will be calculated).

At Hill AFB, four points along the groundwater flow path were chosen for comparison of corrected and observed BTEX concentrations to assess the effects of dispersion, dilution from recharge, volatilization, and sorption (Figure 1a). Table 2 shows corrected BTEX concentrations and the percent of BTEX lost due to biodegradation between these points. The calculations presented in this table confirm that biodegradation of the BTEX compounds is occurring at this site. The specific mechanisms of intrinsic bioremediation operating at Hill AFB are discussed below.

Mechanisms of Intrinsic Bioremediation at Hill AFB

The information presented in the preceding section shows that intrinsic bioremediation of dissolved BTEX is occurring at the POL site. Geochemical data can now be used to determine which mechanisms of intrinsic bioremediation are operating at the site and the relative importance of each mechanism.

General Groundwater Geochemistry. General ambient groundwater geochemistry at the site is conducive to intrinsic bioremediation. Total alkalinity at the site is fairly high, and ranges from 349 to 959 mg/L. This amount of alkalinity is sufficient to buffer potential changes in pH caused by biologically mediated BTEX oxidation reactions. Groundwater pH at the POL facility ranges from 6.3 to 8.3 standard units. This range of pH is optimal for BTEX-degrading microbes. The average temperature of groundwater is 18°C. The reduction/oxidation (redox) potential ranges from 274 to –190 millivolts (mV). Areas at the site with low redox potentials coincide with areas of BTEX contamination; low dissolved oxygen, nitrate, and sulfate concentrations; and elevated ferrous iron and methane concentrations. This suggests that dissolved BTEX at the site is subjected to a variety of biodegradation processes including aerobic respiration, denitrification, iron reduction, sulfate reduction, and methanogenesis.

Aerobic Respiration. Figure 1b shows the distribution of dissolved oxygen in groundwater in July 1994. Comparison of Figures 1a and 1b shows that areas with elevated total BTEX concentrations have depleted dissolved oxygen concentrations. A similar trend was observed in August 1993. This is an indication that aerobic biodegradation of BTEX is occurring at the site. With a background dissolved oxygen concentration of approximately 6 mg/L, the shallow groundwater at this site has the capacity to assimilate 1,900 µg/L of total BTEX, based on the stoichiometry presented in Table 1.

TABLE 2. BTEX lost due to biodegradation, Hill AFB, July 1994.

Compound	Point A (0) Measured Concentration (µg/L)	Point B (137) Measured Concentration (µg/L)	Point B Corrected Concentration[a] (µg/L)	Percent Lost to Biodegradation[b] Between A and B	Point C (305) Measured Concentration (µg/L)	Point C Corrected Concentration (µg/L)	Percent Lost to Biodegradation Between B and C	Point D (335) Measured Concentration (µg/L)	Point D Corrected Concentration (µg/L)	Percent Lost to Biodegradation Between C and D
benzene	5,600	458	2,693	57	7	8	100	1	1	92
toluene	5,870	10	59	99	10	11	0	1	1	95
ethylbenzene	955	454	2,667	0	23	25	100	4	4	97
p-xylene	1,620	272	1,597	2	26	28	99	12	13	90
m-xylene	5,130	442	2,599	54	18	20	100	19	21	0
o-xylene	2,300	51	300	89	3	3	100	6	7	0
Total BTEX	21,475	1,686	9,915	58	87	95	99	40	47	85
trimethylbenzene	2,123	361	2,123	0	330	361	0	295	330	0

(a) See text for calculation of corrected concentration.
(b) Percent lost to biodegradation = $((Measured_A - Corrected_B)/(Measured_A - Measured_B))*100$.

Denitrification. Figure 1c shows the distribution of nitrate + nitrite (as N) in groundwater in July 1994. Comparison of Figures 1a and 1c shows that areas with elevated total BTEX concentrations have depleted nitrate + nitrite concentrations. Comparison of Figures 1b and 1c shows that areas with depleted dissolved oxygen concentrations have depleted nitrate + nitrite concentrations. Similar trends were observed in August 1993. These relationships provide strong evidence that anaerobic biodegradation of the BTEX compounds is occurring at the site through the microbially mediated process of denitrification. Background nitrate + nitrite (as N) concentrations are about 8.0 mg/L, which is equivalent to 29 mg/L of NO_3^-. This nitrate (as NO_3^-) concentration was calculated by assuming that all nitrate + nitrite (as N) is present as nitrate. This assumption is valid because nitrite is metastable in the groundwater environment and nitrite is seldom present in concentrations large enough to influence the ionic balance to a noticeable degree (Hem 1985). This is especially true in uncontaminated (background) areas where denitrification is not occurring. Ionic nitrate and nitrite data collected in January 1994 confirm this. With a background nitrate (as NO_3^-) concentration of approximately 29 mg/L, the shallow groundwater at this site has the capacity to assimilate 6,000 μg/L of total BTEX during denitrification, based on the stoichiometry presented in Table 1.

Iron Reduction. Figure 1d shows the distribution of ferrous iron in groundwater in July 1994. Comparison of Figures 1a and 1d shows that areas with elevated total BTEX concentrations have elevated ferrous iron concentrations. A similar trend was observed in August 1993. This is an indication that ferric iron is being reduced to ferrous iron during biodegradation of BTEX compounds. Background ferrous iron concentrations are less than 0.05 mg/L. The highest measured ferrous iron concentration was 50.5 mg/L. This suggests that the shallow groundwater at this site has the capacity to assimilate 2,300 μg/L of total BTEX during iron reduction, based on the stoichiometry presented in Table 1. This calculation is based on observed ferrous iron concentrations, and not on the amount of ferric hydroxide available in the aquifer. Therefore, potential iron assimilative capacity could be much higher.

Sulfate Reduction. There was no clear relationship between BTEX and sulfate concentrations in August 1993. Figure 1e shows the distribution of sulfate in groundwater in July 1994. Comparison of Figures 1a and 1e shows that by July 1994, areas with elevated total BTEX concentrations had depleted sulfate concentrations. This is an indication that anaerobic biodegradation of the BTEX compounds is occurring at the site through the microbially mediated process of sulfate reduction. With a background sulfate concentration of about 100 mg/L, the shallow groundwater at this site has the capacity to assimilate 21,000 μg/L of total BTEX during sulfate reduction, based on the stoichiometry presented in Table 1.

Methanogenesis. Available geochemical evidence suggests that methano-genesis, like sulfate reduction, may have become a more important BTEX-degradation mechanism between August 1993 and July 1994. There was no clear relationship between BTEX and methane concentrations in August 1993. Figure 1f shows the distribution of methane in groundwater in July 1994. Comparison of Figures 1a and 1f shows that areas with elevated total BTEX concentrations have elevated methane concentrations. This is an indication that anaerobic biodegradation of the BTEX compounds is occurring at the site through the microbially mediated process of methanogenesis with carbon dioxide serving as the electron acceptor. This is consistent with other electron acceptor data collected at the site, with the area having elevated methane con-centrations being confined to areas with depleted dissolved oxygen, nitrate, and sulfate concentrations and elevated ferrous iron concentrations. The high-est measured methane concentration within the BTEX plume was 2.04 mg/L, and background concentrations are less than 0.001 mg/L. This suggests that the shallow groundwater at this site has expressed the capacity to assimilate 2,600 µg/L of total BTEX during methanogenesis, based on the stoichiometry presented in Table 1. These calculations are based on observed methane con-centrations and not on the amount of carbon dioxide available in the aquifer. Therefore, methanogenic assimilative capacity could be much higher.

Total Expressed Assimilative Capacity

The data presented in the preceding sections suggest that mineralization of BTEX compounds is occurring through the microbially mediated processes of aerobic respiration, denitrification, iron reduction, sulfate reduction, and methanogenesis. On the basis of site geochemical data and the stoichiometry presented in Table 1, the expressed BTEX assimilative capacity of groundwater at the POL facility is at least 33,800 µg L (Table 3). The measured concentrations of ferrous iron and methane may not be the maximum expressed or achievable. The highest dissolved BTEX concentration observed at the site was 21,475 µg/L. On the basis of the calculations presented in the preceding sections and obser-vations made at this site over the period from August 1993 through July 1994, groundwater in the vicinity of the POL facility has sufficient assimilative capac-ity to degrade dissolved BTEX that partitions from the LNAPL plume into the groundwater before the plume migrates 500 m downgradient from the source area. The difference between observed assimilative capacity and the highest BTEX concentration probably results from the biodegradation of other petro-leum-related compounds, such as naphthalene, dissolved in the groundwater.

PATTERNS OF INTRINSIC BIOREMEDIATION AT PATRICK AIR FORCE BASE, FLORIDA

An estimated 700 gal (2,659 L) of unleaded gasoline was released into the soil and shallow groundwater at the Base Exchange (BX) Service Station (Site ST-29), Patrick AFB, Florida, in 1986. Several investigative techniques,

TABLE 3. Expressed assimilative capacity of site groundwater.

	Expressed BTEX Assimilative Capacity (µg/L)	
Process	Hill AFB	Patrick AFB
Aerobic Respiration	1,900	1,200
Denitrification	6,000	—
Ferric Hydroxide Reduction	2,300	90
Sulfate Reduction	21,000	—
Methanogenesis	2,600	17,400
Expressed Assimilative Capacity	33,800	18,690
Highest Observed Total BTEX Concentration	21,475	7,304

including soil and groundwater sampling and aquifer testing, were utilized at this site. Cone penetrometer testing and hollow-stem auger drilling were used to collect stratigraphic information and soil samples. Groundwater samples were collected at monitoring points installed in CPT holes and at monitoring wells. Slug tests were conducted in monitoring wells.

Site Geology and Hydrogeology

Patrick AFB lies on a narrow barrier island that parallels the eastern Florida coastline and is bounded on the east by the Atlantic Ocean (230 m east of the site) and on the west by the Banana River (730 m west of the site). The ground surface at Site ST-29 slopes gently westward toward the Banana River.

Shallow subsurface deposits at Site ST-29 consist of fine- to coarse-grained marine sand that is poorly to moderately sorted and contains up to 40% shell fragments. These sand deposits extend to a depth of approximately 7.6 m and contain interspersed organic matter. The Caloosahatchee Marl formation underlies the sands at this site and acts as an aquitard. The water table is present at depths of 1.22 to 1.52 m bgs and is unconfined. Groundwater elevation data indicate that flow is to the west, toward the Banana River and away from a divergent groundwater divide located along the eastern edge of the site. Water level measurements indicate that the local horizontal hydraulic gradient is approximately 0.002 m/m. Vertical hydraulic gradients measured in monitoring point nests range from 0.000 m/m to 0.003 m/m (downward).

Results of slug testing indicate that the average hydraulic conductivity of the shallow saturated zone is approximately 0.026 cm/s. With a gradient of 0.002 m/m, a hydraulic conductivity of 0.026 cm/s, and assuming an effective porosity of 0.35, the average advective groundwater velocity is 0.13 m/day or approximately 48 m/year. On the basis of measured TOC concentrations in uncontaminated areas at the site, a median retardation factor of 2.6 was calculated for benzene. This gives an estimated retarded solute transport velocity of 18.3 m/year.

Nature and Extent of Contamination

Figures 2a through 2d show the distribution of soil contamination, dissolved BTEX, electron acceptors, and metabolic by-products at Patrick AFB, Florida. Mobile LNAPL has not been detected in monitoring wells or monitoring points at Site ST-29. Residual BTEX contamination resulting from vertical and lateral migration of hydrocarbons is found over the area indicated in Figure 2a. The highest observed concentration of residual total BTEX in soil is 1,236 mg/kg. Benzene was detected in this sample at 6.99 mg/kg.

Figure 2a also shows the distribution of total BTEX dissolved in groundwater. Dissolved BTEX contamination is migrating to the west in the direction of groundwater flow. Dissolved BTEX contamination at Patrick AFB is limited to the shallow saturated zone. The maximum dissolved BTEX concentration observed at the site in March 1994 was 7,304 μg/L. It is likely that dissolved BTEX concentrations were much higher shortly after the spill occurred in 1986 [potentially as high as 120 mg/L, based on equilibrium partitioning considerations (American Petroleum Institute 1985)].

Documented Loss of Contaminants at the Field Scale

Three points along the groundwater flow path were chosen to correct observed BTEX concentrations for the effects of dispersion, dilution from recharge, sorption, and volatilization (Figure 2a). Point A was chosen to coincide with the highest observed dissolved BTEX concentration. Points B and C are located 38 m and 98 m, respectively, downgradient of point A.

Table 4 shows TMB-corrected BTEX concentrations and the percent of each BTEX compound lost via biodegradation between Points A and B and B and C. The results of these calculations indicate that TMB is not entirely recalcitrant under the conditions found at this site (benzene concentrations were higher at Point B than at Point A). However, any biodegradation of TMB results in underestimation of the percentage of BTEX biodegraded, so the calculations presented in Table 4 are conservative, and confirm that biodegradation is occurring. The mechanisms of intrinsic bioremediation likely operating at this site are discussed below.

Mechanisms of Intrinsic Bioremediation at Patrick AFB

The information presented in the preceding section shows that intrinsic bioremediation of dissolved BTEX is occurring at Site ST-29. Geochemical data can now be used to determine which mechanisms of intrinsic bioremediation are operating at the site and the relative importance of each mechanism.

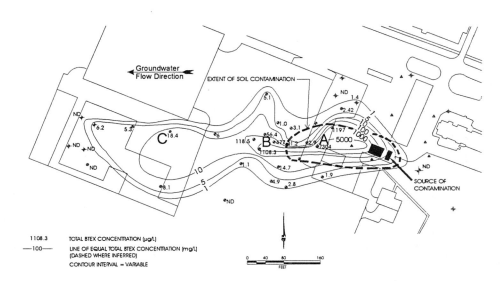

FIGURE 2a. Distribution of total BTEX in groundwater at Patrick AFB, Florida (where 1 ft = 0.3 m).

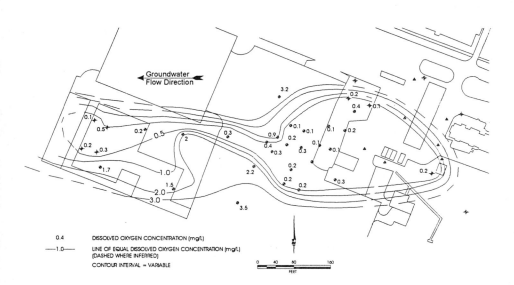

FIGURE 2b. Distribution of dissolved oxygen in groundwater at Patrick AFB, Florida (where 1 ft = 0.3 m).

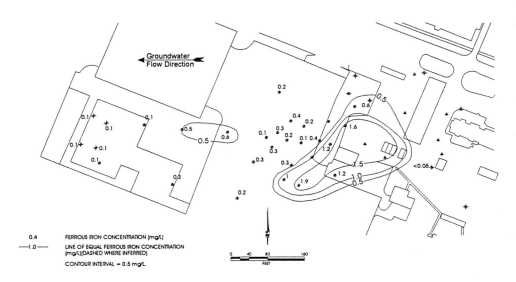

FIGURE 2c. Distribution of ferrous iron in groundwater at Patrick AFB, Florida (where 1 ft = 0.3 m).

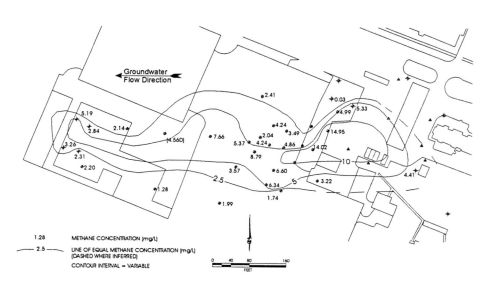

FIGURE 2d. Distribution of methane in groundwater at Patrick AFB, Florida (where 1 ft = 0.3 m).

TABLE 4. BTEX lost due to biodegradation, Patrick AFB, March 1994.

	Location (meters downgradient of highest BTEX concentration)						
Compound	Point A (0) Measured Concentration (µg/L)	Point B (38) Measured Concentration (µg/L)	Point B Corrected Concentration[a] (µg/L)	Percent Lost to Biodegradation[b] Between A and B	Point C (98) Measured Concentration (µg/L)	Point C Corrected Concentration (µg/L)	Percent Lost to Biodegradation Between B and C
benzene	724	960	25,714	0	1	9	99
toluene	737	17	455	39	2	19	0
ethylbenzene	823	12	321	62	2	19	0
p-xylene	1,220	39	1,045	15	4	37	6
m-xylene	2,410	37	991	60	7	65	0
o-xylene	1,390	44	1,179	16	4	37	18
trimethylbenzene	750	28	750	0	3	28	0

(a) See text for calculation of corrected concentration.
(b) Percent lost to biodegradation = $((\text{Measured}_A - \text{Corrected}_B)/(\text{Measured}_A - \text{Measured}_B)) \times 100$.

General Groundwater Geochemistry. The ambient groundwater geochemistry at the site is conducive to intrinsic bioremediation. Total alkalinity at the site varies from 148 mg/L to 520 mg/L and is sufficient to buffer potential changes in pH caused by biologically mediated BTEX oxidation reactions. Groundwater pH at Site ST-29 ranges from 6.7 to 7.6 standard units, well within the optimal range of 6 to 8 for BTEX-degrading microbes. Temperatures in the shallow saturated zone range from 24.7 to 27.8°C. These are relatively high temperatures for shallow groundwater, suggesting that bacterial growth rates could be high. The redox potential at Site ST-29 ranges from 54 mV to –293 mV. Areas of the site with low redox potential coincide with areas of high BTEX contamination, low dissolved oxygen concentrations, slightly elevated ferrous iron concentrations, and significantly elevated methane concentrations. These characteristics suggest that dissolved BTEX at the site is subjected to a variety of biodegradation processes including aerobic respiration, iron reduction, and methanogenesis.

Background nitrate concentrations are extremely low at this site, ranging from less than 0.05 mg/L to 0.29 mg/L. Nitrate reduction cannot be a significant BTEX removal mechanism because sufficient quantities of nitrate are not available for metabolism. Although sulfate concentrations are fairly high at this site (up to 86 mg/L), no relationship between sulfate and BTEX is apparent.

Aerobic Respiration. Figure 2b shows the distribution of dissolved oxygen in groundwater at Site ST-29. Comparison of Figures 2a and 2b shows that areas with elevated total BTEX concentrations coincide with areas having depleted dissolved oxygen concentrations. This is an indication that aerobic biodegradation of the BTEX compounds is occurring at the site. With a background dissolved oxygen concentration of approximately 3.7 mg/L, the shallow groundwater at this site has the capacity to assimilate 1,200 μg/L of total BTEX, based on the stoichiometry presented in Table 1.

Iron Reduction. Figure 2c shows the distribution of ferrous iron in groundwater at Site ST-29. Comparison of Figures 2a and 2c shows that areas with elevated total BTEX concentrations have slightly elevated ferrous iron concentrations. This suggests that ferric iron is being reduced to ferrous iron during biodegradation of BTEX compounds. Background ferrous iron concentrations are about 0.1 mg/L. The highest measured ferrous iron concentration within the BTEX plume was 1.9 mg/L, suggesting that the shallow groundwater at this site has expressed the capacity to assimilate at least 90 μg/L of total BTEX during iron reduction, based on the stoichiometry presented in Table 1. This calculation is based on observed ferrous iron concentrations and not on the amount of ferric hydroxide available in the aquifer. Therefore, iron assimilative capacity could be much higher.

Methanogenesis. Figure 2d shows the distribution of methane in groundwater. Comparison of Figures 2a and 2d shows that areas with elevated total

BTEX concentrations correlate with elevated methane concentrations. This is an indication that anaerobic biodegradation of the BTEX compounds is occurring at the site. This is consistent with other electron acceptor and redox potential data for this site. The highest measured methane concentration was 14.6 mg/L. Background concentrations of methane are about 1 mg/L. This suggests that the shallow groundwater at this site has expressed the capacity to assimilate at least 17,400 μg/L of total BTEX during methanogenesis, based on the stoichiometry presented in Table 1. These calculations are based on observed methane concentrations and not on the amount of carbon dioxide available in the aquifer. Therefore, methanogenic assimilative capacity could be much higher.

Total Expressed Assimilative Capacity. The data presented in the preceding sections suggest that mineralization of BTEX compounds is occurring through the microbially mediated processes of aerobic respiration, iron reduction, and methanogenesis. On the basis of site geochemical data and the stoichiometry presented in Table 1, the expressed BTEX assimilative capacity of groundwater at Site ST-29 is at least 18,690 μg/L (Table 3). The measured concentrations of ferrous iron and methane may not be the maximum achievable. The highest dissolved BTEX concentration observed at the site was 7,304 μg/L. On the basis of the calculations presented in the preceding sections, and on site observations, groundwater at Site ST-29 has sufficient assimilative capacity to degrade dissolved BTEX that partitions from the residual phase in soil into the groundwater before the plume migrates 400 m downgradient from the source area.

CONCLUSIONS

The prevailing mechanisms of natural attenuation and intrinsic bioremediation site depend on the governing physical and chemical characteristics of the shallow subsurface. Two lines of evidence were used to document the occurrence of intrinsic bioremediation in this paper: documented loss of contaminants at the field scale, and geochemical evidence. Field and geochemical data indicate that intrinsic bioremediation is occurring at both Hill AFB and Patrick AFB. Available geochemical data suggest that patterns of intrinsic bioremediation are different at each of these Air Force bases. However, anaerobic processes account for the greatest mass of BTEX destroyed at both sites. This pattern is typical of most of the sites studied by the authors to date using the technical protocol developed by the Air Force Center for Environmental Excellence (Wiedemeier et al. 1994).

Historical data and geochemical evidence show that intrinsic bioremediation of dissolved BTEX is occurring at the POL facility at Hill AFB. Geochemical evidence suggests that aerobic respiration, denitrification, ferric iron reduction, sulfate reduction, and methanogenesis are the primary processes contributing to microbial degradation of BTEX at Hill AFB. The distribution of electron acceptors and metabolic byproducts relative to total BTEX concentrations changed noticeably between August 1993 and July 1994. In August 1993, there was good a correlation between areas with depleted dissolved oxygen, depleted nitrate, elevated ferrous iron, and elevated BTEX concentrations. No correlations between sulfate, methane, and BTEX concentrations were apparent at this time. By July 1994, however, areas with depleted sulfate and elevated methane concentrations exhibited an excellent correlation with areas containing elevated BTEX concentrations. This suggests that sulfate reduction and methanogenesis are becoming more important microbial degradation mechanisms as the plume matures. As indicated by expressed assimilative capacity, sulfate reduction is now the dominant mechanism of BTEX biodegradation at this site (Table 3). Vroblesky and Chapelle (1994) show that the dominant terminal electron accepting processes in an aquifer with fuel hydrocarbon contamination can vary both temporally and spatially due to consumption, recharge, and migration of electron acceptors.

Geochemical data show that intrinsic bioremediation of dissolved BTEX is occurring at Site ST-29 at Patrick AFB. These data suggest that aerobic respiration and methanogenesis are the primary processes contributing to microbial degradation of BTEX at Patrick AFB. Reduction of ferric iron also appears to play a limited role in microbial degradation of BTEX at this site. There is good correlation between areas with depleted dissolved oxygen, elevated ferrous iron, elevated methane, and elevated BTEX concentrations at the site. No correlation between nitrate, sulfate, and BTEX concentrations was apparent in March 1994. In contrast to the site at Hill AFB, it appears that aerobic respiration at the periphery of the plume and methanogenesis in the anaerobic core of the plume are the primary microbiologically mediated BTEX oxidation processes operating at Patrick AFB. As indicated by expressed assimilative capacity, methanogenesis is the dominant mechanism of BTEX biodegradation at this site (Table 3).

REFERENCES

American Petroleum Institute. 1985. *Laboratory Study on Solubilities of Petroleum Hydrocarbons in Groundwater.* API Publication Number 4395.

Bouwer, E.J. 1992. "Bioremediation of Subsurface Contaminants." In R. Mitchell (Ed.), *Environmental Microbiology.* Wiley-Liss, New York, NY.

Buscheck, T.E. and C.M. Alcantar. 1995. "Regression Techniques and Analytical Solutions to Demonstrate Intrinsic Bioremediation." In R.E. Hinchee, J.T. Wilson, and D.C. Downey (Eds.), *Intrinsic Bioremediation,* pp. 109-116. Battelle Press.

Hem, J.D. 1985. *Study and Interpretation of the Chemical Characteristics of Natural Water.* United States Geological Survey Water Supply Paper Number 2254, 263 p.

National Research Council. 1993. *In Situ Bioremediation, When does it Work?* National Academy Press, Washington, DC.

Vroblesky, D.A., and F.H. Chapelle. 1994. "Temporal and spatial changes of terminal elec-tron-accepting processes in a petroleum hydrocarbon-contaminated aquifer and the sig-nificance for contaminant biodegradation." *Water Resources Research,* *30*(5)1561-1570.

Wiedemeier, T.H., D.C. Downey, J.T. Wilson, D.H. Kampbell, R.N. Miller, and J.E. Hansen. 1994. Technical Protocol for Implementing the Intrinsic Remediation with Long-Term Monitoring Option for Natural Attenuation of Fuel Contamination Dissolved in Groundwater (Draft). Air Force Center for Environmental Excellence, Brooks Air Force Base, TX.

Wiedemeier, T.H., M.A. Swanson, R.T. Herrington, J.T. Wilson, D.H. Kampbell, R.N. Miller, and J.E. Hansen. 1995. Comparison of Two Methods for Determining Biodegradation Rate Constants at the Field Scale. In Preparation.

Simulation of Natural Attenuation with Multiple Electron Acceptors

Hanadi S. Rifai, Charles J. Newell, Ross N. Miller,
Sam Taffinder, and Mark Rounsaville

ABSTRACT

Natural attenuation has recently emerged as a potential remedial alternative at contaminated sites. Groundwater transport and fate models are generally used to assess the viability of using natural attenuation at a site. These models have traditionally focused on modeling advection, dispersion, and sorption—three of the main attenuation mechanisms in groundwater. A fourth key variable that impacts natural attenuation is biodegradation. The BIOPLUME II model is one of the few public domain models that simulate aerobic biodegradation in groundwater. This paper describes a major enhancement to the model, referred to as BIOPLUME III. The enhanced model simulates aerobic and anaerobic processes due to the presence of multiple electron acceptors: oxygen, nitrate, iron, sulfate, and carbon dioxide. The BIO-PLUME III model simulates sequential biodegradation processes depending on the availability of electron acceptors in the aquifer.

INTRODUCTION

Intrinsic bioremediation is a management strategy that relies on physical, chemical, and biological processes in an aquifer to attenuate contaminant concentrations to acceptable levels. In order for the viability of intrinsic bioremediation to be assessed at a site, detailed site characterization and data analysis must be conducted to quantify the extent of natural attenuation. Additionally, and because intrinsic bioremediation has to be protective of public health and the environment, exposure and risk-assessment analysis is also required. This means that quantitative analysis of fate and transport at a naturally attenuating site is necessary. Although many analytical and numerical models exist today, and most of these models simulate advection, dispersion, sorption, and source dissolution mechanisms in contaminated aquifers, few simulate biological transformations.

One of these models, BIOPLUME II, has been widely used for simulating natural attenuation by the federal and private sectors. The BIOPLUME II model was developed at Rice University in the late 1980s. The model was specifically designed to simulate aerobic biodegradation, with the assumption that the transport of oxygen into the contaminated aquifer is the limiting process for biodegradation. This work describes a major enhancement of the BIOPLUME II model. The enhanced model, referred to as BIOPLUME III, simulates biodegradation at a site due to the presence of multiple electron acceptors: oxygen, nitrate, iron, sulfate, and carbon dioxide. BIOPLUME III allows a variety of kinetic expressions for biodegradation for each individual electron acceptor. More importantly, however, the model allows simulation of sequential utilization of a number of electron acceptors at a site. These added capabilities allow users to simulate aerobic biodegradation in aquifer zones where oxygen exists, and anaerobic biodegradation in zones where oxygen is absent.

OVERVIEW OF THE BIOPLUME II MODEL

The BIOPLUME II model was developed in the late 1980s by Rifai et al. (1988). The model was based on the U.S. Geological Survey (USGS) two-dimensional method-of-characteristics (MOC) program. The USGS MOC model simulates flow and transport of contaminants in groundwater, and incorporates the processes of advection, dispersion, and sorption. Rifai et al. (1988) modified the MOC model to incorporate biodegradation.

Two expressions for biodegradation were simulated: (1) first-order decay of the contaminants, and (2) aerobic decay based on the dissolved oxygen concentrations present in groundwater. To simulate the aerobic biodegradation of contaminants, Rifai et al. (1988) relied on earlier research completed by Borden and Bedient (1986), which showed that when biodegradation occurs rapidly relative to groundwater velocities, the process can be assumed to occur instantaneously. In other words, the rate of the reaction can be neglected, and the biodegradation of contaminants using oxygen as an electron acceptor is based solely on the stoichiometry of the chemical reaction.

The instantaneous reaction assumption allowed Rifai et al. (1988) to simulate the transport of two components with the BIOPLUME II model, those being hydrocarbon and oxygen. The oxygen and hydrocarbon are allowed to react using a superposition assumption (Figure 1). Oxygen is depleted in areas where hydrocarbon concentrations are elevated. Hydrocarbon concentrations are depleted by an amount that is stoichiometrically equivalent to the supply of oxygen that is present at the given location.

The BIOPLUME II model incorporated three different sources of oxygen: (1) naturally occurring oxygen in groundwater prior to contamination (this is essentially equivalent to the background concentration of oxygen in the aquifer); (2) oxygen supply through injection wells (i.e., bioremediation); and (3) dissolved oxygen in the flowing groundwater. A fourth source of oxygen,

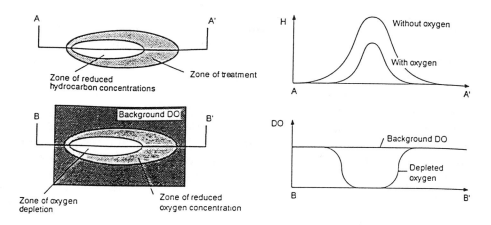

FIGURE 1. Superposition principle used in BIOPLUME II.

namely reaeration or diffusion of oxygen from the unsaturated zone into the saturated zone, was simulated indirectly by assuming that its impact on groundwater can be represented as a first-order decline in contaminant concentrations.

Rifai et al. (1988) completed detailed sensitivity analyses using the BIOPLUME II model. Their results indicated that aerobic biodegradation was most sensitive to the hydraulic conductivity and the reaeration coefficient. Aerobic biodegradation was independent of sorption, and not very sensitive to dispersion.

The BIOPLUME II model, since its development, has become widely accepted for simulating aerobic biodegradation and enhanced bioremediation using oxygen as a terminal electron acceptor. A number of researchers have used the model in simulating these processes at their sites (Chiang et al. 1989; Rifai et al. 1988; Caldwell et al, 1992; Wiedemeier et al. 1995). The model adequately simulated the aerobic processes at these sites, with the exception that the model allows total depletion of oxygen at a given location, a phenomenon that is not supported by field evidence. In general, field data indicate that there is an oxygen concentration threshold below which aerobic biodegradation will not occur.

Despite its popularity, the BIOPLUME II model has two main shortcomings: (1) it does not allow for simulating of slowly degrading compounds under aerobic conditions (i. e., compounds that do not conform to the instantaneous reaction assumption); and (2) it does not allow for simulating anaerobic processes affected by other electron acceptors (the model does allow for a first-order decay mechanism).

Rifai and Bedient (1990) extended the BIOPLUME II model to incorporate aerobic biodegradation using Monod kinetics. They demonstrated that the instantaneous reaction assumption was adequate for aquifers and contaminants that resulted in a large Damkohler number. The Damkohler number is the ratio of the rate of the reaction to the velocity of the groundwater.

The need to simulate anaerobic processes in groundwater is the main motivation behind the present work. Much interest has been generated in using natural attenuation as a remedial alternative at sites that do not pose a significant risk to human health and the environment. As such, modeling tools are needed that adequately represent the biological reactions occurring at a site. Recent research indicates that anaerobic biodegradation of petroleum hydrocarbons and chlorinated aliphatics may account for a significant portion of the observed mass loss in a contaminant plume.

THE BIOPLUME III MODEL

Aerobic and Anaerobic Biodegradation

The BIOPLUME III model, in contrast with BIOPLUME II, simulates the transport and fate of six components in groundwater. These are the contaminant, dissolved oxygen, nitrate, iron, sulfate, and carbon dioxide. Each of these components, except for the contaminant, is either present as a background characteristic of the groundwater or can be introduced into groundwater using injection wells.

The biodegradation model that is assumed in BIOPLUME III is one of sequential utilization of electron acceptors. As an example, the aerobic and anaerobic reactions for benzene are shown in Table 1. At any given location in the model grid, the program will initiate a biodegradation sequence based on the availability of electron acceptors at that location and assuming the following order in the occurrence of the reactions:

$$O_2 \Rightarrow NO_3^- \Rightarrow Fe^{3+}\ SO_4^{2-} \Rightarrow CO_2$$

TABLE 1. Aerobic and anaerobic biodegradation of benzene.

Degradation Type	Reaction Mechanism
Aerobic:	$C_6H_6 + 7.5O_2 \Rightarrow 6CO_2 + 3H_2O$
Denitrification:	$C_6H_6 + 6H^+ + 6NO_3^- \Rightarrow 6CO_2 + 3N_2 + 6H_2O$
Iron Reducing:	$C_6H_6 + 30Fe^{3+} + 12H_2O \Rightarrow 6CO_2 + 30H^+ + 30Fe^{2+}$
Sulfate Reducing:	$C_6H_6 + 3.75SO_4^{2+} + 7.5H^+ \Rightarrow 6CO_2 + 3.75H_2S + 3H_2O$
Methanogenesis:	$C_6H_6 + 4.5H_2O \Rightarrow 2.25CO_2 + 3.75CH_4$

In BIOPLUME III, oxygen is used first and, unless the oxygen concentration decreases to below a user-defined threshold at a given location, anaerobic processes will not take place.

For each of the electron acceptors, a number of biodegradation kinetic expressions can be selected by the user as follows:

1. Aerobic biodegradation: first-order decay, instantaneous or Monod kinetics.
2. Anaerobic biodegradation using nitrate: first-order decay, instantaneous or Monod kinetics.
3. Anaerobic biodegradation for all other electron acceptors: first-order decay or Monod kinetics.

Therefore, the user can mix a number of kinetic expressions at a site depending on the electron acceptor being used. The type of kinetic expression selected by the user can be determined from laboratory or field studies. For example, Monod kinetic data for the anaerobic electron acceptors can be determined from laboratory experiments.

Other Enhancements to the Model

One of the significant physical processes that impacts the spread of a contaminant plume is the dissolution of contaminants from residual sources present in the aquifer. These sources include free-phase chemicals and soil contamination above or below the water table. In BIOPLUME II, two types of contaminant sources can be simulated: (1) a constant concentration source, and (2) a constant flux source. Constant flux sources are usually modeled using injection wells with a specified injection rate and source concentration. The user has to estimate the injection rate and source concentration based on the characteristics of the source remaining in the aquifer.

Recent research on residual sources, however, indicates that the behavior of residual sources may not be adequately described using a continuous source as modeled in BIOPLUME II. Rather, dissolution mechanisms may be better simulated as a decaying source. The BIOPLUME III model will account for this type of source. It is noted, however, that dissolution from free-phase liquids (as in the case of light and dense nonaqueous-phase liquids) cannot be simulated using this type of expression. Other expressions are being considered at this time.

It should be noted that the BIOPLUME III model, like its predecessor BIOPLUME II, treats the organic as a single compound or lumped parameter. At this point, our knowledge of the fundamental processes does not justify modeling the organic on a compound by compound basis.

User Interfaces

The BIOPLUME II model was developed to run on an IBM-compatible personal computer (PC). An interactive menu-driven data-input utility was incorporated into the model. Results from the program could be translated to provide output files that could be displayed using contouring software such as the SURFER program. A Macintosh version of the BIOPLUME II model was also developed at that time. The Macintosh version uses graphical input and output interfaces instead of their menu-driven counterparts on the PC.

Over the last few years since the BIOPLUME II model was developed, computer technology has experienced a tremendous revolution in terms of speed and performance. The development of more sophisticated interfaces for BIOPLUME III is thus feasible. The BIOPLUME III model will run in both a Windows-based environment and on a Macintosh platform. Additionally, both platforms will use state-of-the-art spreadsheet and graphical interfaces that allow visual display and analysis of model results.

The BIOPLUME III model, like BIOPLUME II, will be in the public domain and will be distributed through Rice University. The model will be available in the summer 1995.

Field Applications

In a companion field effort to the model development, several field sites are being sampled to collect contamination data and multiple electron acceptor information. These data currently are being used to calibrate and verify BIO-PLUME III.

REFERENCES

Borden, R. C., and P. B. Bedient. 1986. "Transport of Dissolved Hydrocarbons Influenced By Oxygen-Limited Biodegradation: 1. Theoretical Development." *Water Resources Research* 13:1973-1982.

Caldwell, K. R., D. L. Tarbox, K. D. Barr, S. Fiorenza, L. E. Dunlap, and S. B. Thomas. 1992. "Assessment of Natural Bioremediation as an Alternative to Traditional Active Remediation at Selected Amoco Oil Company Sites, Florida." In *Proceedings of the 1992 Petroleum Hydrocarbons and Organic Chemicals in Ground Water: Prevention, Detection, and Restoration*, pp. 509-525. Water Well Journal Publishing Co., Dublin, OH.

Chiang, C. Y., J. P. Salanitro, E. Y. Chai, J. D. Colthart, and C. L. Klein. 1989. "Aerobic Biodegradation of Benzene, Toluene, and Xylene in a Sandy Aquifer: Data Analysis and Computer Modeling." *Ground Water* 27: 823-834.

Rifai, H. S., and P. B. Bedient. 1990. "Comparison of Biodegradation Kinetics With an Instantaneous Reaction Model." *Water Resources Research* 26(4)637-645.

Rifai, H. S., P. B. Bedient, J. T. Wilson, K. M. Miller, and J. M. Armstrong. 1988. "Biodegradation Modeling at an Aviation Fuel Spill Site." *ASCE Journal of Environmental Engineering* 114 (5):1007-1029.

Wiedemeier, T. H., M. A. Swanson, J. T. Wilson, D. H. Kampbell, R. N. Miller, and J. E. Hansen. 1995. "Patterns of Intrinsic Bioremediation at Two United States Air Force Bases." Submitted to *Proceedings of the 1995 Petroleum Hydrocarbons and Organic Chemicals in Ground Water: Prevention, Detection, and Restoration.* Water Well Journal Publishing Co., Dublin, OH.

Taking Advantage of Natural Biodegradation

William A. Butler and Craig L. Bartlett

ABSTRACT

A chemical manufacturing facility in central New Jersey evaluated alternatives to address low levels of volatile organic compounds (VOCs) in groundwater. Significant natural attenuation of VOCs was observed in groundwater, and is believed to be the result of natural biodegradation, commonly referred to as intrinsic bioremediation. A study consisting of groundwater sampling and analysis, field monitoring, and transport modeling was conducted to evaluate and confirm this phenomenon. The primary conclusion that can be drawn from the study is that observed natural attenuation of VOCs in groundwater is due to natural biodegradation. Based on the concept that natural biodegradation will minimize contaminant migration, bioventing has been implemented to remove the vadose-zone source of VOCs to groundwater. Taking advantage of natural biodegradation has resulted in significant cost savings compared to implementing a conventional groundwater pump-and-treat system, while still protecting human health and the environment.

INTRODUCTION

A chemical manufacturing facility in central New Jersey evaluated alternatives to address low levels of VOCs in groundwater. The remediation objective is to meet applicable state groundwater quality standards at the site perimeter. A conventional approach to meet this objective would consist of a 350-gal/min (1,325-L/min) groundwater pump-and-treat system. This approach is costly and may not be necessary since there is no immediate risk to potential off-site receptors; therefore, innovative alternatives were evaluated.

Groundwater analytical data indicate that significant natural attenuation of VOCs is occurring. This observed attenuation is believed to be the result of naturally occurring biodegradation, commonly referred to as intrinsic bioremediation. If natural biodegradation is occurring, a source-focused alternative, complemented by natural biodegradation, will meet the remediation objective more cost effectively than a conventional pump-and-treat system. A study consisting of groundwater sampling and analysis, field monitoring, and transport modeling was conducted to evaluate and confirm this phenomenon. This paper summarizes the results and conclusions of the study.

SITE HYDROGEOLOGY

The site is located in a recharge area of a regional aquifer in central New Jersey. The aquifer is unconfined and consists of fine- to medium-grained sand with discontinuous beds of silty clay, and is underlain by a continuous bed of low-permeability clay. Depths to water range from 20 to 80 ft (6 to 24 m) because of topographical relief, and saturated thickness ranges from 40 to 80 ft (12 to 24 m). Hydraulic conductivity is 9.8×10^{-4} ft/s (3×10^{-2} cm/s) and hydraulic gradient is 0.004 ft/ft. Assuming a 20% porosity, groundwater velocities are estimated to be 2 ft/day (0.6 m/day).

SAMPLING AND ANALYSIS

VOCs (including benzene, toluene, ethylbenzene, xylenes, acetone, and ketones) have been detected in groundwater. Two adjacent former tank farms were determined to be the primary source of the VOC plume, based on significant concentrations of VOCs in vadose-zone soils and light, nonaqueous-phase liquids (LNAPLs) detected on the water table in these areas. In groundwater beneath the tank farms, concentrations of individual VOCs have been observed at concentrations ranging from 1 to 100 mg/L, and are significantly less downgradient toward the site perimeter.

Groundwater samples were collected from selected monitoring wells as part of this study and analyzed for VOCs and other water-quality parameters. United States Environmental Protection Agency's (U.S. EPA's) analytical method 524.2 was selected for VOCs because of the low detection limits achievable by this method. Table 1 presents the average VOC concentrations in groundwater at the source area and downgradient intervals. Current and past analytical data indicate that significant natural attenuation of VOCs is occurring. VOC concentrations are high at the source and several orders of magnitude lower to nondetectable at the site perimeter.

Table 1 also presents average concentrations of other parameters at the source area and downgradient intervals. The VOC-indicator parameters, including biochemical oxygen demand (BOD), chemical oxygen demand (COD), and total organic carbon (TOC), also exhibit significant natural attenuation. Correlations between BOD, COD, and VOC concentrations indicate that the majority of the VOCs present are biodegradable. If groundwater exerted a COD, but no BOD, attenuation of VOCs could not be explained by natural biodegradation.

TABLE 1. Average groundwater concentrations (μg/L) versus downgradient distance.

Parameter	Source	Downgradient Distance 0 - 500 ft	500 - 1,000 ft	1,000 - 2,500 ft
Benzene	2,600	30	NA[a]	ND
Toluene	46,700	5,500	1	1
Ethylbenzene	7,100	270	ND	ND
Xylenes	25,300	1,100	ND	1
Acetone	21,700	ND	ND	ND
MEK[b]	1,700	10	ND	ND
MIBK[c]	1,600	3	ND	ND
Dissolved oxygen[d]	100	1,000	4,000	6,500
Iron (Fe^{3+} and Fe^{2+})	60,000	50,000	10,000	2,000
Iron (Fe^{2+})	50,000	50,000	6,000	ND
Sulfate	240,000	70,000	NA[e]	NA
BOD	350,000	30,000	1,000	1,000
COD	360,000	60,000	20,000	30,000
TOC	60,000	8,000	2,000	1,000

(a) ND (not detected).
(b) MEK (methyl ethyl ketone).
(c) MIBK (methyl isobutyl ketone).
(d) Dissolved oxygen concentrations approximated from Figure 1.
(e) NA (not analyzed).

The presence of dissolved or reduced iron (Fe^{2+}) indicates that microorganisms may be using iron (Fe^{3+}) as an electron acceptor. When oxygen or other electron acceptors with higher oxidation-reduction potential are not available, anaerobic and/or facultative microorganisms will use Fe^{3+} as an electron acceptor and reduce it to Fe^{2+} (McAllister and Chiang 1994). Sulfate is present and can also be used by microorganisms as an electron acceptor to degrade VOCs (McAllister and Chiang 1994).

FIELD MONITORING

Field monitoring consisted of performing in situ measurements of dissolved oxygen (DO) in groundwater throughout the site, using a multifunctional probe. Figure 1 is an isoconcentration map of DO that depicts an area of groundwater containing low DO levels (< 1 mg/L) relative to surrounding groundwater. Groundwater in this area contains high VOC concentrations and corresponds to the former tank farms determined to be the primary VOC source to groundwater. Average DO concentration in this area is 0.8 mg/L versus 4.3 mg/L in surrounding groundwater outside the VOC plume.

A localized area of high VOC concentrations and low DO, surrounded by areas of low VOC concentrations and high DO, strongly suggests that indigenous microorganisms are using oxygen in the source area to degrade the VOCs

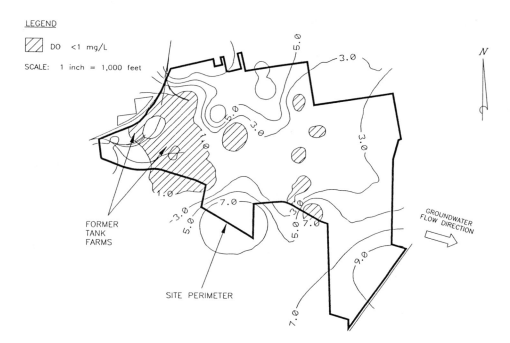

FIGURE 1. Dissolved oxygen concentrations (mg/L).

present (Salanitro 1993; McAllister and Chiang 1994). Groundwater BOD data further support this conclusion.

TRANSPORT MODELING

Transport modeling can help demonstrate natural biodegradation (McAllister and Chiang 1994). For this study, groundwater transport modeling was conducted to estimate the biodegradation rate of VOCs in groundwater, compare results to reported literature values, and determine the relative effect of dispersion and sorption alone. Benzene was used as the representative VOC for the modeling.

Analytical model PLUME2D (IGWMC 1986) was selected to model benzene transport. This model solves the advection dispersion groundwater transport equation for a continuous point source based on the solution given by Wilson and Miller (1978). The model requires input parameters of groundwater velocity, aquifer thickness, longitudinal and transverse dispersivity, retardation factor, source mass loading rate and duration of loading, and first-order decay constant. Retardation was estimated to be 1.7 based on site-specific TOC measurements of aquifer material, and K_{oc} values from literature. Time of source loading was assumed to be 15 years based on site history.

In order to fit model results to observed data, the source loading rate, dispersivity (longitudinal and transverse), retardation, and benzene decay rate were varied. A first-order decay mechanism was used to estimate benzene biodegradation. Although this is not strictly appropriate to simulate oxygen-limited biodegradation, this method was assumed to be sufficient to verify that attenuation was occurring beyond what would be expected from other mechanisms such as dispersion and retardation. The source loading rate was adjusted to ensure that the benzene concentration at the source was always 5.8 mg/L (maximum concentration at source).

A longitudinal dispersivity of 300 m (984 ft) and a transverse dispersivity of 12 m (39 ft) were used for modeling. The longitudinal dispersivity value was estimated based on U.S. EPA literature (U.S. EPA 1990). The ratio of longitudinal to transverse dispersivity is velocity dependent, and a value of 25 was used, which is approximately equal to that presented by Freeze and Cherry (1979) at an average linear velocity of 0.6 m/day (2 ft/day). A benzene biodegradation half-life of 90 days was determined to provide the best fit to observed benzene concentrations and attenuation, and is within the range of values presented by Dragun (1988) of between 48 and 110 days.

Sensitivity analyses were performed on parameters of velocity, dispersivity, retardation, aquifer thickness, and porosity using a reasonable range of site-specific values. Modeling results indicate dispersion and sorption cannot be responsible for the observed attenuation without biodegradation. A first-order decay of benzene at a rate consistent with biodegradation laboratory studies approximates the observed plume configuration. Figure 2 is a graph depicting

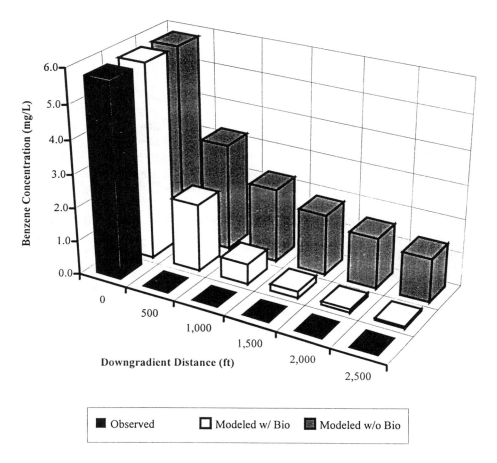

FIGURE 2. Benzene concentration versus downgradient distance.

the modeling results, and compares observed benzene concentrations with concentrations modeled with and without biodegradation at selected down-gradient distances.

CONCLUSIONS

The primary conclusion drawn from this study is that the observed natural attenuation of VOCs in groundwater is due to natural biodegradation, commonly referred to as intrinsic bioremediation. This is supported by the following:

1. Groundwater analytical results show that significant natural attenuation of VOCs is occurring. Concentrations are high at the source area and significantly lower to nondetectable at the downgradient site perimeter.

2. The correlation between BOD, COD, and VOC concentrations indicates that the majority of VOCs present are biodegradable.
3. The presence of potential electron acceptors, including iron (Fe^{3+} and Fe^{2+}) and sulfate, indicates that microorganisms may be using one or more of these compounds as electron acceptors in the source area, in the absence of oxygen.
4. A localized area of low DO and high VOC concentrations, surrounded by areas of higher DO and low VOC concentrations, strongly suggests that indigenous microorganisms are using oxygen in the source area to degrade the VOCs present.
5. Transport modeling indicates the observed attenuation of VOCs in site groundwater is primarily due to natural biodegradation. Modeling results indicate that dispersion and sorption cannot be responsible for the observed attenuation without biodegradation. The plume would be much larger than observed if no biodegradation was occurring.

Confirming that natural biodegradation is occurring in site groundwater has enabled source-focused remedial alternatives to be considered. Based on the concept that natural biodegradation will minimize contaminant migration, bioventing has been implemented to remove the vadose-zone source. Taking advantage of natural biodegradation has resulted in significant cost savings, compared to implementing a perimeter groundwater pump-and-treat system, while still achieving the remediation objective and being protective of human health and the environment.

REFERENCES

Dragun, J. 1988. *The Soil Chemistry of Hazardous Materials.* Hazardous Materials Control Research Institute, Silver Spring, MD.

Freeze, R.A., and J.A. Cherry. 1979. *Groundwater.* Prentice-Hall, Inc., Englewood Cliffs, NJ.

International Ground Water Modeling Center (IGWMC). 1986. *PLUME2D.* Golden, CO.

McAllister, P.M., and C.Y. Chiang. 1994. "A Practical Approach to Evaluating Natural Attenuation of Contaminants in Groundwater." *Ground Water Monitoring and Remediation*: pp. 161-173.

Salanitro, J.P. 1993. "The Role of Bioattenuation in the Management of Aromatic Hydrocarbon Plumes in Aquifers." *Ground Water Monitoring and Remediation*: pp. 150-161.

U.S. Environmental Protection Agency. 1990. *A Subtitle D Landfill Application Manual for the Multimedia Exposure Assessment Model (MULTIMED).*

Wilson, J.L., and P.J. Miller. 1978. "Two-Dimensional Plume in Uniform Groundwater Flow." *Journal of the Hydraulics Division 104* (HY4). American Society of Civil Engineers.

Enhanced Aerobic Bioremediation of Residual Hydrocarbon Sources

Paul M. McAllister, Chen Yu Chiang, Joseph P. Salanitro, Ira J. Dortch, and Patty Williams

ABSTRACT

Column studies were conducted with subsurface soils to evaluate the effect of oxygen enhancement on in situ bioremediation of residual hydrocarbon sources. A mixture of benzene, toluene, and *m*-xylene (BTX) in decane was injected into two packed-soil columns. A synthetic groundwater was passed through the columns at one to three pore volumes per day. One column received a flow of nitrogen purged groundwater. The feed to the other column was oxygenated to 10 to 20 mg/L dissolved oxygen. Results of column and microcosm experiments demonstrated that biodegradation and consumption of dissolved oxygen did occur in the zone of the residual hydrocarbon source. However, the net impact on the leaching of BTX was not significant. Soluble BTX concentrations sampled at a port just downgradient of the source revealed no substantial differences between the behavior of both columns and an equilibrium dissolution model. Our data suggest that most of the dissolved oxygen fed may have been used for biodegradation of the residual decane.

INTRODUCTION

The effectiveness of enhanced aerobic biodegradation has been clearly demonstrated for soluble plumes of benzene, toluene, ethylbenzene, and xylenes (BTEX) (Chiang et al. 1993, Salanitro 1993). However, the effects of enhanced bioremediation on residual hydrocarbons present in the saturated zone have not been demonstrated. Trapped residual hydrocarbons are often the long-term source of groundwater contamination, and the time required to remediate the source dictates the overall remediation time for a site. Thus, the objective of this work was to conduct laboratory studies (batch microcosms and columns) to determine if increased dissolved-oxygen (DO) levels in groundwater would significantly enhance remediation of the residual source zone.

EXPERIMENTAL APPARATUS
AND PROCEDURES

To evaluate the effects of enhanced aerobic biodegradation on a residual hydrocarbon source, a column study was conducted. One soil column was fed a synthetic groundwater with DO levels of 10 to 20 mg/L (the high-DO column), and a second column received a feed of nitrogen-purged water with a DO < 1 mg/L (the low-DO column). The 2.54-cm-i.d. × 60.96-cm glass columns had eight sampling ports at 7.62-cm intervals along the length of each column. The columns were packed wet with a medium Michigan sand (90% sand, 2% silt, 8% clay). Caps filled with glass beads were placed on both ends of the columns and flowthrough cells were placed at the inlet and outlet for continuous monitoring of DO concentrations. Soil porosities were 0.28 and 0.26 for the low- and high-DO columns, respectively. The hydraulic conductivities were 0.020 cm/s and 0.021 cm/s for the low- and high-DO columns, respectively. An Orion Model 840 DO meter and probe continuously read and recorded the DO at the inlet of the low-DO column. Nonoxygen-consuming Leeds and Northrup model 7932 meters, and biochemical oxygen demand (BOD) were used to record DO continuously at the inlet and outlet of the high-DO column.

After assembly and characterization of the columns, 10.5 ml of a hydrocarbon mixture composed of 0.25% benzene, 1.25% toluene, 1.25% *m*-xylene, and 97.25% decane was injected through port 8 of each column from a syringe pump. Syringe needles were inserted through Teflon™ diaphragms at each of the sampling ports and left in place throughout the experiment. Samples were drawn by opening valves placed on the ends of the syringe needles. These samples were analyzed by purge-and-trap gas chromatography (Varian Model 3700, Tekmar Model LSC2000 purge and trap, and Supelco Carbopack B with 3% SP mobile phase) for BTX, or small amounts of water were bled from the column through the syringe needles into a flow-through cell in which an oxygen microelectrode (Microelectrodes, Inc. OM-4 Meter and 16-730 Probe) was placed. A synthetic groundwater was prepared based on the mineral composition of the site groundwater. It had the following composition: 24 mg/L Mg^{++}, 71 mg/L Cl^-, 78 mg/L Ca^{++}, 5 mg/L SO_4^{2-}, and 2 mg/L Na^+. The feed to each column was amended with NH_4Cl to give 0.9 mg/L N, and K_2HPO_4 to give 0.05 mg/L P. The feedwater was held in two separate 60-L tanks and infiltrated upward through the columns using peristaltic pumps. The feed for the low-DO column was purged with nitrogen and held with a nitrogen blanket in the feed tank at 0.3 psi. The feed to the high-DO column was purged with a 50/50 O_2/N_2 mixture and held under this gas mixture at 0.3 psi. Groundwater feed to the low-DO column was held consistently at < 0.1 mg/L DO. The inlet DO to the high-DO column was less stable, and varied from 8 to 20 mg/L throughout the experiment.

COLUMN STUDIES

During operation of the columns, the water flowrate, DO, and BTX concentrations at nine points from the inlet to the outlet of the columns were determined. The flowrate and dissolved oxygen data are shown in Figure 1. Throughout the first 33 days of the experiment, the flowrate to each column was maintained at approximately 0.011 cm³/s (~3 pore volumes/day). During this initial period, the consumption of DO in the high-DO column was minimal. The inlet DO was held relatively constant between 10 and 15 mg/L, with the outlet DO varying from 8 to 13 mg/L. Thus, DO was not consumed instantaneously in the column. After 33 days, the flowrate was decreased to approximately 0.0039 cm³/s (~1 pore volume/day). Within 2 days of changing the flowrate, the outlet DO decreased to < 1 mg/L. The change in flowrate had no impact on the feed to the low-DO column as the inlet and outlet DO were maintained at < 0.1 mg/L throughout the experiment.

Although measurement of the inlet- and outlet-DO concentrations showed that nearly complete oxygen consumption was occurring, it did not determine where along the length of the column the oxygen was consumed. DO profiles along the length of the high-DO column are shown in Figure 2 at two times during the experiment. It is important to note that the residual hydrocarbon was injected into the column approximately 7.6 cm above the bottom of the column, and was contained within a 5- to 10-cm zone above the injection point. It can be seen in Figure 2 that nearly all the DO was consumed in the source area. Low-DO concentrations (< 1 mg/L) remained to be used for biodegradation in the upper 50 cm of the column. Under these conditions, instantaneous

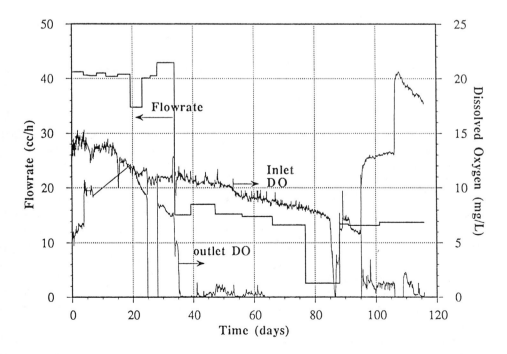

FIGURE 1. Flowrate and inlet and outlet DO for the high-DO column.

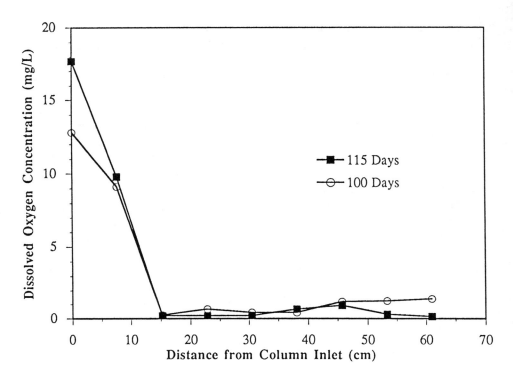

FIGURE 2. Axial dissolved oxygen profiles in the high-DO column.

consumption of DO in the source zone was apparent, but it was not clear if DO was used for degradation of BTX or decane.

Additional data collected during the course of the experiment were the concentrations of BTX at each of the sampling ports. To evaluate the impact of enhanced bioremediation on the residual source zone, data from port 5, which was 22.9 cm downgradient of the hydrocarbon injection point, are shown in Figure 3(a) for benzene and *m*-xylene, and Figure 3(b) for toluene. The solid lines in Figure 3 represent the predicted equilibrium dissolution curves for the columns (Rixey et al. 1991). The time is rendered nondimensional in Figure 3 by multiplying it by the water flowrate and dividing by the volume of hydrocarbon injected to the column. The concentrations of each compound decreased in a manner consistent with the equilibrium dissolution curves down to approximately 10 to 100 μg/L, where concentrations began to fluctuate. As expected, benzene decreased first because it was the most soluble component, followed by toluene, and then *m*-xylene. The BTX data from port 5 of each column alone did not show significant differences between the low- and high-DO columns.

Monitoring of inlet and outlet DO concentrations, sampling of BTX, monitoring of flowrate, and postexperiment sectioning of the column made it possible

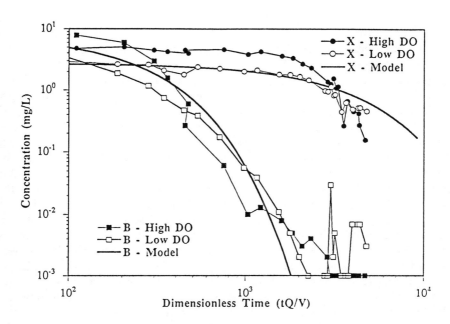

FIGURE 3(a).   Comparison of benzene and *m*-xylene concentrations in the low- and high-DO columns at port 5.

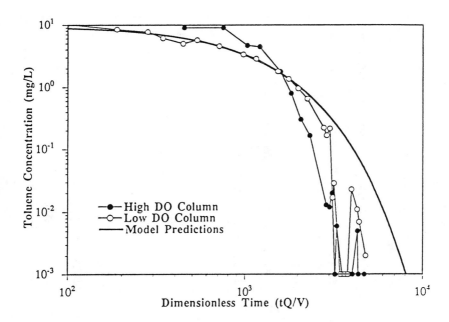

FIGURE 3(b).   Comparison of toluene concentrations in the low- and high-DO columns at port 5.

to conduct mass-balance analyses on DO and BTX. The residual amounts of BTX and decane were determined by sectioning the columns at 2-inch (5.08-cm) intervals, extracting a 10-g sample from each section with methanol followed by gas chromatography-mass spectrometry analysis. Figure 4 shows the mass-balance data for benzene, toluene, and *m*-xylene for the high- and low-DO columns. Shown are the initial mass (0.019 mg, 0.096 mg, and 0.096 mg, of benzene, toluene, and *m*-xylene respectively), the mass flushed from the columns in the soluble phase, the residual detected on the soil at the end of the experiment, and the remainder, which must be accounted for by biodegradation, physical loss, or experimental and analytical error. In terms of total recovery, 71% and 65% of the BTX was recovered by flushing from the low- and high-DO columns, respectively. Residual BTX was 1.% and 0.2%, respectively, for the low- and high-DO columns. For both columns, dissolution and flushing was the major mechanism of removal. The difference in recovery was 6.8% of the total BTX injected to the columns. Thus, oxygen enhancement resulted in degradation of, at most, 6.8% of the BTX in the high-DO column. No significant difference was observed for benzene between the two columns, but the data shown in Figure 4 suggest some oxygen was utilized to degrade soluble toluene and *m*-xylene. Figures 3(a) and (b) suggest, however, that this biodegradation had no discernible impact on soluble concentrations near the source.

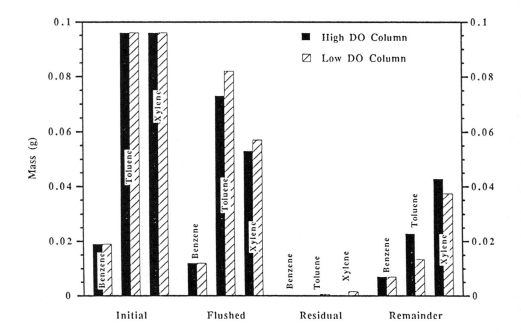

FIGURE 4. Final mass balance of benzene, toluene, and xylene in high-DO and low-DO columns.

After 115 days, 684 mg of oxygen had been fed to the high-DO column and by difference across the column, 518 mg were consumed. Assuming that all the BTX was converted to carbon dioxide and water, 518 mg of oxygen was sufficient to degrade 167 mg of BTX, which was 80% of the total BTX (211 mg) originally placed in the column. As indicated above, only 6.8% or 14 mg of BTX in the high-DO column could have been aerobically biodegraded. Thus, another sink for oxygen was present. Consumption of DO to degrade a portion of the 7.45 g of residual decane injected into the column is a likely explanation, as indicated by the microcosm study results which follow. Of the 7.45 g of decane present, the DO supplied was sufficient to mineralize only 0.148 g ($\sim$ 2%) after 170 pore volumes of water at 10 to 20 mg/L DO.

MICROCOSM STUDIES

Soil-slurry microcosms were established to determine the presence of BTX-degrading microbes, and to evaluate the impact of residual separate-phase decane on soluble-phase BTX biodegradation. Microcosms were prepared by aseptically transferring 90 g wet weight of soil to 250-mL serum bottles with Teflon™-lined septa and aluminum crimp seals. The synthetic groundwater was adjusted to pH 8.25 and DO of 15 mg/L before addition to the microcosms. Standard mixtures of BTX were prepared in water (200 mg/L benzene, 400 mg/L toluene, and 100 mg/L m-xylene), and decane (0.25 wt% benzene, 1.25 wt% toluene, and 1.25 wt% m-xylene). BTX solutions in water or decane were added to microcosms at high, medium, or low levels corresponding to 15 to 20, 1.2 to 1.5, and 0.1 to 0.15 mg/L total BTX, respectively. Respiration-inhibited controls were amended with 2% sodium azide to inhibit soil bacterial activity, and all microcosms were incubated at 20 to 23°C. At multiple times over 21 days, 2-mL aqueous samples were removed from the serum bottles with a sterile needle and syringe, preserved with 6N HCl, and then analyzed by purge-and-trap gas chromatography as described for the column samples.

Estimates of the initial number of bacteria in the soil were 10^6 to 10^7 heterotrophs and 10^4 to 10^7 BTX degraders per g of wet soil. The results of the microcosm studies are shown in Table 1. The microcosm studies show that (1) aromatic hydrocarbons were rapidly degraded in < 14 days at 20 to 23°C in soil-groundwater slurries containing water soluble compounds at < 1500 μg/L total BTX and 200 to 350 mg/L benzene, and high DO of 15 mg/L; (2) the calculated rates of BTX biodegradation in slurries with or without mineral nutrients (NH_4^+, PO_4^{3-}) were two to five times higher when hydrocarbons were added as water-soluble components than as a residual phase in decane (however, degradation did occur in the presence of decane); and (3) addition of nutrients stimulated microbial degradation of hydrocarbons at medium (< 1500 μg/L) but not high (15,000 μg/L) levels of BTX. These experiments suggest that aromatic compounds biodegrade at slower rates when a large, rapidly degradable residual hydrocarbon such as decane is present. In this case, microbial oxygen consumption and oxidation of decane may reduce the available electron acceptor

TABLE 1. Results of microcosm studies for biodegradation in the presence of decane.

BTX Solvent	BTX Concentrations (mg/L)	Nutrient Addition	First-Order BTX Decay Rate (day⁻¹)
Water	High—(13 to 20)	–	0.03, 0.05
		+	0.03, 0.09
	Medium —(1.2 to 1.5)	–	0.48, 0.61
		+	2.40
	Low—(0.1 to 0.2)	–	0.48, 0.61
		+	2.40
Decane	High—(13 to 20)	–	0.02, 0.04
		+	0.02, 0.03
	Medium —(1.2 to 1.5)	–	0.14, 0.26
		+	0.39, 0.55
	Low—(0.1 to 0.2)	–	0.46, 0.54
		+	0.66, 0.74

(DO) required for the removal of dissolved-phase BTX. Also, the continued solubilization of BTX from a residual hydrocarbon phase would influence the disappearance of soluble BTX.

SUMMARY

This study indicated that although active biodegradation occurred in the presence of residual decane, the net impact on the leaching of BTX was not significant. Dissolution and flushing were the major mechanisms (>70%) responsible for removal of BTX from the residual source. Significant consumption of dissolved oxygen occurred rapidly in the source area and was most likely consumed for biodegradation of decane. The results suggest that after infiltrating high-DO (10 to 20 mg/L) groundwater for 4 months, only a small fraction (~ 2% of the decane) of the applied hydrocarbon may have degraded. While biodegradation did occur in the source zone of the high-DO column, the overall dissolution and leaching of BTX from the source were not significantly affected.

REFERENCES

Chiang, C. Y., P. D. Petkovsky, M. C. Beltz, S. J. Rouse, and T. Boyd. 1993. "An enhanced bioremediation system at a central production facility: System design and data analysis." *Proceedings of the Conference on Petroleum Hydrocarbons and Organic Chemicals in Ground Water: Prevention, Detection, and Restoration,* Nov. 10-12, Houston, TX, pp. 661-678.

Rixey, W. G., P. C. Johnson, G. M. Deeley, D. L. Byers, and I. J. Dortch. 1991. "Mechanisms for the removal of residual hydrocarbons from soils by water, solvent, and surfactant flushing." In E. J. Calabrese and P. T. Kostecki (Eds.), *Hydrocarbon Contaminated Soils*, Vol. 1, pp. 387-409. Lewis Publishers, Inc., Chelsea, MI.

Salanitro, J. P. 1993. "An industry's perspective on intrinsic bioremediation." Presented at the National Research Council (Water Science & Technology Board) Committee on "In Situ Bioremediation: When Does It Work?," Washington DC, October 26-29, 1992.

Postremediation Bioremediation

Richard A. Brown, Patrick M. Hicks,
Ronald J. Hicks, and Maureen C. Leahy

ABSTRACT

In applying remediation technology, an important question is when to stop operations. Conventional wisdom states that each site has a limit of treatability. Beyond a point, the site conditions limit access to residual contaminants and, therefore, treatment effectiveness. In the treatment of petroleum hydrocarbons, the issue in ceasing remedial operations is not what is the limit of treatment, but what should be the limit of effort. Because hydrocarbons are inherently biodegradable, there is a point in remediation where natural or intrinsic bioremediation is adequate to complete the remedial process. This point is reached when the rate of residual carbon release is the limiting factor, not the rate of oxygen or nutrient supply. At such a point, the rate and degree of remediation is the same whether an active system is being applied or whether nothing is being actively done. This paper presents data from several bioremediation projects where active remediation was terminated above the desired closure levels. These site data illustrate that intrinsic bioremediation is as effective in site closure as continued active remediation.

INTRODUCTION

Two questions are frequently asked about the application of remedial technology. First, what levels of treatment need to be or can be achieved; and second, how long will it take to achieve an end point (i.e., when to cease treatment).

Although the answer to the question of what needs to be achieved often is readily apparent, the question of what end point can, in fact, be achieved is much more difficult to answer. The most honest answer is WYGIWYG ("what you get is what you get"), the environmental equivalent of the computer parlance WYSIWYG ("what you see is what you get"). WYGIWYG is often expressed as "limits of performance" or "asymptotic performance." Although cleanup goals are often set, they are not always met.

The answer to the question of how long it will take to achieve an end point is that the more rigorous the cleanup standards, the longer the projected operational time. An industrial-use cleanup standard of ~1 mg/L total volatile organic compounds (VOCs) may elicit an operational time estimate of 3 to 5 years, whereas the goal of drinking water standards on the order of 1 to 5 μg/L for specific compounds yields estimates of 10 to 20 years, i.e., the colloquialism for "forever." If either the desired end point cannot be achieved or the end point will take 10 to 20 years of effort, there is a reluctance to employ rigorous and expensive technology. Even more discouraging to the use of innovative remedial technology is the specter of having to use a completely different technology, such as excavation and incineration, if the original technology cannot achieve the end point.

However, for many contaminants there is an alternative answer to both of these questions: postremediation bioremediation, which is the use of intrinsic bioremediation as the last stage of site remediation. Intrinsic bioremediation uses naturally occurring biological processes to attenuate organic contaminants. For contaminants that are readily biodegradable, postremediation (intrinsic) bioremediation offers a means to both limit the time of active remediation and attain low-containment-level cleanup goals.

Closure is often defined as attaining cleanup standards uniformly across the site. However, what typically limits closure is access to the residual contaminants by the treatment system. In the final stages of remediation the treatment system is often governed by diffusive flow rather than advective flow and access to the residual contaminants is physically limited and slow. Postremediation bioremediation addresses this lack of treatment efficiency by using naturally occurring biological processes to effectively treat residual contaminants without the costly operation of a full-scale remedial system. The key to successful postremediation bioremediation is reducing the carbon level to the point that intrinsic bioremediation can attain the desired level without the continued expenditure of significant effort. Application of postremediation bioremediation thus requires determining the point at which sufficient remediation has been achieved so that intrinsic bioremediation can be maintained. This generally is the point at which the available electron acceptors are in balance with the residual carbon.

Intrinsic bioremediation can use a wide range of electron acceptors. These are listed in Table 1 (Wilson & Armstrong 1994, Wilson et al. 1994, Borden et al. 1994). The table lists the electron acceptors in order of increasing reducing conditions. Of these electron acceptors, oxygen and carbon dioxide are the most readily available because they are replenished by natural recharge processes. Sulfate, iron, and manganese also occur naturally but are dependent on site mineralogy. The predominant sources of nitrite are anthropogenic activities such as agricultural fertilization. Which of these electron acceptors is used is dependent on the redox conditions (aerobic vs. anaerobic) in the subsurface. In intrinsic bioremediation oxygen is consumed first; then, once anaerobic conditions have been established, nitrate, sulfate, and the other electron acceptors

TABLE 1. Common naturally occurring electron acceptors.

Electron Acceptor	Metabolic Pathway	Natural Sources
Oxygen (O_2)	Aerobic	Atmosphere
Nitrate/Nitrite (NO_3^-/NO_2^-)	Facultative Anaerobic	Agriculture/Septic/ *Atmosphere*[a]
Sulfate (SO_4^{2-})	Anaerobic	Minerals/*Atmosphere*[a]
Iron (Fe^{+3})	Anaerobic	Minerals
Manganese (Mn^{+4})	Anaerobic	Minerals
Carbon Dioxide (CO_2)	Methanogenic	Atmosphere/*Minerals*[a]

(a) Secondary source (italics).

are consumed in order of their redox potentials. Oxygen depletion requires the presence of a carbon substrate that is readily amenable to aerobic biodegradation (e.g., hydrocarbons). Carbon sources that are resistant to all but methanogenic processes (e.g., polychlorinated biphenyls [PCBs]) will not drive the depletion of oxygen, and thus will not "activate" other pathways.

THE CASE FOR POSTREMEDIATION BIOREMEDIATION

In the application of bioremediation, there are generally two stages. In the first stage, the rate-limiting factor is the rate of supply of the electron acceptor. In the second stage of bioremediation, the rate-limiting factor is the availability of the contaminant. In the second stage the system is carbon limited, and intrinsic bioremediation can effectively replace an active bioremediation system.

The availability of the contaminant may be limited by three types of interacting factors; i.e., chemical, structural, and physical, as summarized in Table 2. The lower the permeability, or the higher the heterogeneity, or the higher the K_{OC}, the more contaminant will be sorbed to the soil and not readily available for biodegradation. In such situations, the site may be carbon limited at a relatively high degree of contamination. In these cases, the effectiveness of conventional bioremediation, which is predicated on the addition of electron acceptors and nutrients, is substantially reduced compared to cases with more favorable lithological characteristics. Even though there may be high levels of contaminants on the site, they are not readily accessible to the influx of nutrients and electron acceptors.

Such sites are good candidates for postremediation bioremediation. When a site becomes carbon limited, the contaminant that leaches into the ground-

TABLE 2. Factors limiting contaminant availability.

Type	Category	Impact
Chemical	Sorption (K_{OC})	Solubility/Leaching
Chemical	Soil Concentration	Partitioning/Leaching
Structural	Fractures	Access
Physical	Heterogeneity	Access
Physical	Permeability	Flowrate/Access

water is readily degraded by ambient processes. The unassisted influx of electron acceptors is more than adequate to degrade the carbon that is infused into the system by the leaching of residual contamination. Thus, intrinsic bioremediation is effective in removing the residual contaminants, as illustrated by the following two bioremediation projects involving gasoline contamination.

At the first site, groundwater was contaminated by gasoline from a leaking underground storage tank with initial concentrations exceeding 15 mg/L total petroleum hydrocarbons (TPHs). The geology of the site consisted of interbedded highly fractured red shales and pebbly sandstones to a depth of 25 to 30 ft, over a more competent fine-grained sandstone bedrock. Depth to water was 30 to 35 ft, near the top of the competent bedrock.

The initial remedial system consisted of a centrally located recovery well. The recovered groundwater was treated by air stripping, amended with nutrients, and reinjected through an infiltration gallery located in the original tank pit where the leak had occurred. Supplemental oxygen was added to the site through a series of six air sparging wells equipped with porous-stone diffusers connected to air compressors.

During approximately 2.5 years of operation, the average level of contamination across the site was reduced by approximately 78%, from 4.5 mg/L to 1.1 mg/L, as shown in Figure 1. Groundwater concentrations became static after 1.5 to 2 years, indicating that the rate of biodegradation was limited. Dissolved-oxygen (DO) levels remained lower than 1 to 2 mg/L in the monitoring wells. The system was then modified to increase the rate of oxygenation by adding hydrogen peroxide at ~500 mg/L continuously through the infiltration gallery and on a batch basis through individual monitoring wells. DO levels increased across the site from 1 to 2 mg/L to 3 to 4 mg/L. The modified system was operated for an additional 1.5 years. The results were a drop within 3 to 4 months in contaminant levels from 1.1 to 0.5 mg/L, a 55% decrease. Over another 6 months of operation, the contaminant level dropped to ~200 to 300 μg/L and stabilized. DO levels began to rise after ~15 months of operation, indicating that the rate of biodegradation was slowing. Because no further benefit was being attained and the cost of continued operation was

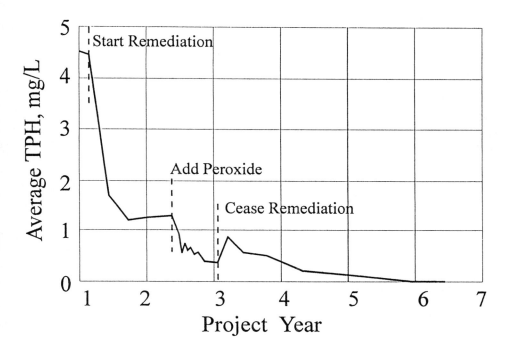

FIGURE 1. Case A. Decrease in TPH with time.

unacceptable, the remedial system was shut down and the site was monitored on a quarterly basis (Figure 1). As is frequently observed, contaminant levels rebounded after system shutdown. After the rebound, a steady decline in contaminant levels from the rebound level of ~900 μg/L was observed to below the detection limit of 50 μg/L after only 5 years of monitoring. This continued drop in contaminant levels after the system was shut down can be attributed to intrinsic bioremediation.

At the second site, groundwater had also been contaminated by a leaking underground storage tank over approximately 15 acres, with concentrations exceeding 50 mg/L benzene, toluene, ethylbenzene, and xylenes (BTEX). The geology of the site consisted of 15 to 20 ft (4.6 to 6 m) of a sand and silt glacial till, underlain by a highly weathered shale bedrock. Depth to water varied from 15 to 30 ft (4.6 to 9 m) across the site. The initial remedial system consisted of three recovery wells. The recovered water was treated by air stripping, amended with nutrients, and reinfiltrated through two infiltration galleries. Supplemental oxygen was sparged into 12 individual wells using air compressors that maintained DO levels of ~1-2 mg/L.

The results, averaged for monitoring wells located within 50 ft of the site (source wells) and for downgradient monitoring wells located to the west and east of the site (plume wells), are depicted in Figure 2. Source wells showed an

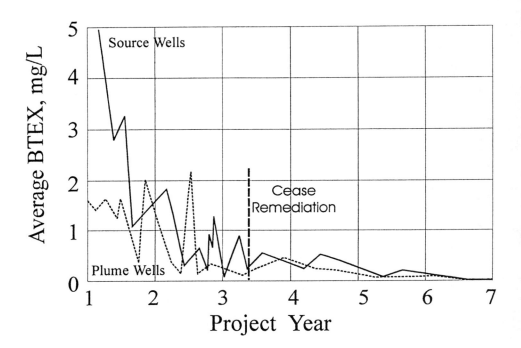

FIGURE 2. Case B. Decrease in BTEX with time.

80 to 90% drop in BTEX over a 3-year period of system operation. The plume wells showed a 70 to 80% drop in BTEX. Because of the residual gasoline trapped in the fractures, BTEX persisted across the site with substantial fluctuations in concentration. BTEX levels increased when the water table dropped. When the water table was higher and the water flow was in the more permeable till, which had been remediated, the BTEX levels dropped. Active bioremediation was terminated and a long-term groundwater recovery system using only one recovery well was operated. The site was monitored for an additional 3 years. BTEX levels in the source wells and in the plume wells continued to drop without further active treatment. Because the release of contaminants was structurally limited, the site was carbon-limited. Continued remediation by intrinsic bioremediation was as effective as active bioremediation at a fraction of the cost.

THE APPLICATION OF POSTREMEDIATION BIOREMEDIATION

Postremediation bioremediation can be a cost-effective means of continuing treatment of degradable contaminants. The process is predicated on the use of intrinsic bioremediation. The principle of operation is to reduce the contaminant load on a site to the point at which intrinsic bioremediation can be sustained and active treatment can be terminated. This point is reached when the site becomes carbon limited; that is, when the rate of remediation is limited by the release of the contaminant from the soil matrix into the treatment stream(s) (air and/or water).

An important question, therefore, is when does a site become carbon limited? For contaminants such as hydrocarbons, which are readily amenable to aerobic biodegradation, the point at which carbon becomes rate limiting can be determined by monitoring oxygen levels. When DO levels can be stabilized above 2 mg/L (Salanitro 1994) and/or gaseous oxygen levels in the vadose zone can be maintained above 5% (Ely & Heffner 1988) under ambient conditions, oxygen is no longer rate limiting as the electron acceptor. Under these conditions, the rate of oxygen utilization due to carbon release from the soil is less than the rate of oxygen replenishment, and these ambient oxygen levels can be maintained naturally without continued operation of an active bioremediation system. (If other electron acceptors are used, their static concentrations should be stable or increasing.)

Figure 3 provides a decision matrix for applying postremediation bioremediation using aerobic degradation. When the rate of mass removal attained by

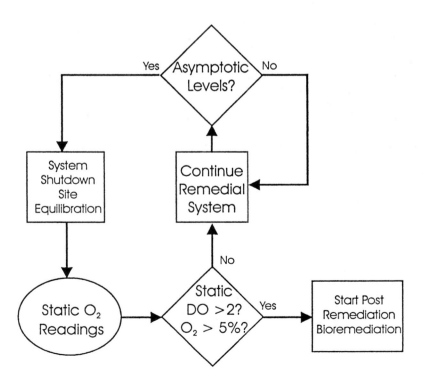

FIGURE 3. Decision matrix for postremediation bioremediation.

the treatment system has significantly slowed, and contaminant concentrations are approaching asymptotic levels, the site may be evaluated for postremediation bioremediation. To determine if postremediation bioremediation is applicable, static DO and/or O_2 measurements need to be taken. To take a static DO/O_2 level, all remedial systems are shut down and the site is allowed to remain static for 7 to 10 days. If the site can maintain target oxygen levels (i.e., 2 mg/L DO, or 5% O_2) under static conditions, intrinsic bioremediation can be sustained and the site can be switched to postremediation bioremediation.

Once postremediation (intrinsic) bioremediation is applied, the site needs to be monitored to determine performance. If the target contaminants show a stable or decreasing concentration, the site is being adequately treated. Oxygen (or other electron-acceptor) levels also should be monitored. If the levels drop below the target values, it may indicate an influx of contaminant(s), which may require restarting the active treatment system.

REFERENCES

Borden, R. C., C. A. Gomez, and M. T. Becker. 1994. "Natural Bioremediation of a Gasoline Spill." In R. E. Hinchee, B. C. Alleman, R. E. Hoeppel, and R. N. Miller (Eds.), *Hydrocarbon Bioremediation*, Lewis Publishers, Boca Raton, FL, pp. 290–295.

Ely, D. L., and D. A. Heffner. 1988. "Process for In-Situ Biodegradation of Hydrocarbon Contaminated Soil." U.S. Patent Number 4,765,902.

Salanitro, J. P. 1994. "An Industry's Perspective on Intrinsic Bioremediation." *In Situ Bioremediation, When Does it Work?* National Academy Press, Washington, DC, pp. 104–109.

Wilson, J. T., and J. Armstrong. 1994. "Traverse City: Geochemistry and Intrinsic Bioremediation of BTEX Compounds." Paper presented at Symposium on Intrinsic Bioremediation of Ground Water, Denver, CO. August.

Wilson, J. T., D. H. Kampbell, and J. Armstrong. 1994. "Natural Bioreclamation of Alkylbenzenes (BTEX) From a Gasoline Spill in Methanogenic Groundwater." In R. E. Hinchee, B. C. Alleman, R. E. Hoeppel, and R. N. Miller (Eds.), *Hydrocarbon Bioremediation*, Lewis Publishers, Boca Raton, FL, pp. 201–218.

Evaluating Natural Attenuation of Petroleum Hydrocarbon Spills

Marleen A. Troy, Katherine H. Baker, and Diane S. Herson

ABSTRACT

Many pump-and-treat operations for remediating petroleum hydrocarbon (PHC)-contaminated groundwater (and soil) have been in operation for several years. Some systems are at the point where the groundwater is no longer impacted, but there is still evidence that pockets of residual PHC contamination remain. The current regulatory environment will not allow these types of sites to be closed and treatment operations terminated. Recently, natural attenuation has been proposed as an alternative for sites where continued operation of groundwater pump-and-treat systems would be of no additional benefit. Natural attenuation is defined as the process of allowing indigenous microorganisms to degrade contaminants that have been discharged to the subsurface without intervention. One of the requirements for implementing natural attenuation is documenting the activity of the indigenous microorganisms and subsequent biodegradation of the contaminants. This documentation is obtained through investigations performed in the field or on site-specific samples of soil and groundwater. However, pump-and-treat operations performed at many sites did not routinely require collecting samples and performing appropriate analyses. An approach to assessing natural attenuation from an ongoing evaluation is discussed.

INTRODUCTION

Groundwater and soil can become contaminated with PHCs as a result of leaking underground storage tanks, broken pipelines, and accidental surface discharges. Pump-and-treat (i.e., the extraction of contaminated water and its treatment aboveground) has been used for the past several years to contain and/or remediate sites contaminated in this way (Mercer et al. 1990). After the contaminated water is extracted, it is typically treated by air stripping and/or with granular activated carbon before discharge to a sewage treatment plant, or reinjection to the subsurface. Operational data from pump-and-treat systems has indicated that although hydraulic control can be maintained this type of operation is not always a viable remedial alternative. Projections from the data suggest that millions of gallons of water will have to be pumped for long periods of time (e.g., 50 years) to achieve the desired cleanup results. While alternative technologies such as air sparging and soil vapor extraction are being implemented at many sites, numerous pump-and-treat systems have been in place for several years (2 to 5 years). Many of these systems may have achieved as much effective treatment as possible, and are now being evaluated to determine if operations can be terminated.

Natural attenuation (i.e., the process of allowing naturally occurring microorganisms to degrade contaminants that have been released into the subsurface) is being recommended for implementation after pump-and-treat systems have reduced the potential for offsite migration of the contaminants (National Research Council 1993). Natural attenuation is not a "No-Action" alternative because it requires documentation of the role of the native microorganisms in eliminating the contaminants through tests performed at field sites or on site-derived samples of soil and groundwater (National Research Council 1993).

Pump-and-treat operations performed at many sites do not routinely require the collection of samples and performance of appropriate analyses that would document the potential for natural attenuation to occur. Many of these systems are now at the point where they have been in operation for several years, and only isolated areas of the site have significant groundwater impact. The continued operation of these systems would have no benefit on the cleanup of these isolated impacted areas. Is the contamination at these sites still a threat and, if so, how can the sites be economically remediated in a timely fashion? Natural attenuation is being proposed as a remedial alternative for these types of situations. As discussed above, the use of natural remediation requires documentation that site conditions are conducive for the indigenous subsurface microorganisms. The question then arises: How much information is required to support the premise that conditions at a site are conducive to natural attenuation? The following case study will be used to evaluate potential answers to this question.

CASE STUDY

Data from the pump-and-treat operation for remediation of a service-station gasoline spill, which occurred as a result of a leaking underground storage tank, will be discussed. The objectives of the pump-and-treat system were to mitigate the spread of the plume hydraulically, and to treat the recovered impacted water aboveground with an air stripper. The contamination was discovered as a result of a site assessment. No information is available to determine when the release of gasoline occurred or how much had leaked.

The site is located in a mixed commercial and residential area. The residential areas are located upgradient of the site, and the commercial areas are located downgradient of the site. No potable-water-supply wells are located within a half-mile radius of the site. The nearest surface-water body is a river located approximately 400 ft downgradient of the site. Soils at the site comprise fine-grained material with a hydraulic conductivity of 10^{-1} gpd/ft^2. Groundwater flows at a gradient of 0.01 ft/ft in the direction of the river. No proximal sensitive receptors were identified.

To track the gasoline plume migration, a series of monitoring wells were installed. These wells were used to monitor benzene, toluene, ethylbenzene, and xylenes (BTEX). Depth to water, temperature and pH were also determined. Because of the small size of the site and economic considerations, sampling was performed on a quarterly basis. After 2 years of operation of the pump-and-treat system, and the recovery and treatment of approximately 2 million gallons of groundwater, the system was shut off. The decision to terminate operation of the pump-and-treat system was made because only one groundwater-monitoring well on site had levels of dissolved BTEX above the designated compliance levels. This decision was agreed upon by the governing regulatory agency, provided a natural attenuation monitoring program was performed.

The site is currently being evaluated to assess if it is a candidate for closure based on natural attenuation. For natural attenuation to be applicable, the rate of BTEX degradation needs to be faster than the rate of contaminant migration (Borden 1993). Migration of the existing BTEX plume will be a function of the natural groundwater-flow velocity. The presence of organic carbon in the subsurface soils will also serve to retard the movement of the BTEX plume, potentially slowing the rate. A groundwater-monitoring program is currently in place to ensure that the plume does not regenerate, and that the dissolved BTEX remains localized and is undergoing biodegradation. Data collected for 16 months since the termination of the pump-and-treat system have indicated that the plume has not regenerated.

Table 1 lists the results from the first three rounds of sampling of the impacted well and a downgradient sentinel well. The results are similar in terms of many of the parameters measured (temperature, pH, redox, iron, nitrate, etc.), but the two microbiologically influenced parameters—carbon dioxide and dissolved oxygen—show pronounced differences. Carbon dioxide levels are significantly increased in the impacted well, and dissolved oxygen is decreased compared to the sentinel well. These results are consistent with increased microbiological activity in the impacted well and may indicate that biodegradation of the contaminants is taking place under indigenous site conditions, i.e. that natural attenuation is occurring. Biogenic gases, however, can be produced by microbial metabolism of any organic compound, not just the contaminants. Thus, the increased levels of these compounds seen in the impacted well may indicate microbial activity, but not necessarily removal of BTEX. Therefore, while these compounds are good preliminary indicators that

TABLE 1. Results from the first three rounds of natural attenuation monitoring.

Location	Date	Temp. (°F)	pH (SU)	Carbon Dioxide (mg/L)	DO (mg/L)	Redox (mV)	Total Iron (mg/L)	Dis. Iron (mg/L)	Nitrate-N (mg/L)	Sulfate (mg/L)	TOC (mg/L)	BTEX (µg/L)	Total Heter. (CFU/mL)	BTEX Degraders (CFU/mL)
Impacted Well	Aug-94	64	7.9	134.0	1.0	180	3.78	0.40	<0.20	2.1	4.6	1,066	NA	NA
	Nov-94	57	7.5	177.0	0.6	81	3.07	0.91	<0.20	5.1	5.0	1,103	1.00e+04	3.74e+03
	Jan-95	43	7.7	98.1	0.8	152	5.17	0.16	0.22	5.4	3.9	21	NA	NA
Sentinel Well	Aug-94	62	8.3	63.7	3.4	158	2.37	0.12	<0.20	24.5	<0.50	<1.0	NA	NA
	Nov-94	57	7.7	57.9	2.4	104	1.06	<0.05	<0.20	23.0	1.7	<1.0	4.58e+04	3.00e+02
	Jan-95	41	7.5	35.0	4.5	168	9.74	0.55	0.28	32.2	1.8	<1.0	NA	NA

Temp. = Temperature
BTEX = Benzene, toluene, ethylbenzene, and xylenes
DO = dissolved oxygen
Redox = oxidation-reduction potential
Tot. Iron = total iron
Dis. Iron = dissolved iron
Nitrate = nitrate as N

TOC = total organic carbon
Total Heter. = total heterotrophic bacteria
CFU/mL = colony-forming units per milliliter
NA = not analyzed

natural attenuation may be feasible, the presence of increased levels of carbon dioxide and decreased levels of dissolved oxygen do not absolutely demonstrate that natural attenuation should be considered for a particular site.

Enumeration of microbial populations can be used to provide additional supportive information when natural attenuation is being considered (Table 1). Enumeration of total heterotrophic bacteria and BTEX-degrading bacteria (Gerhardt et al. 1994) can indicate enrichment in the impacted well by organisms with the appropriate metabolic potential to degrade BTEX. More important than the absolute number of microorganism present is the relative number of each type of microorganism present at the impacted area compared to the sentinel area (Baker & Herson 1994). The ratios of microbial numbers between the sentinel and impacted areas reflect enrichment of the indigenous microbial community by microorganisms with the desired metabolic capabilities. Furthermore, the ratios allow for an estimation of the contribution of these organisms to the overall microbial activity, as reflected in the changes in respiratory gases. The number of total heterotrophs is slightly less in the impacted well compared with the sentinel well. The number of BTEX degraders in the impacted well, on the other hand, is significantly higher compared with the sentinel well. This pattern of little or no change in the number of total heterotrophs and the dramatic increase in the number of BTEX degraders indicates enrichment of the microbial community with the desired microorganisms and supports the implementation of natural attenuation.

Ideally, proof of the feasibility of natural attenuation is possible when site-specific data on microbial numbers and activity, as well as degradation kinetics, are available over significant time periods for a particular site. These data, when available, can then be compared to the rate of migration of the contaminant, which can be obtained using any of several accepted hydrogeologic models. Unfortunately, for many of the pump-and-treat sites currently in operation or nearing the end of their useful life, such historical data are not available. In these instances, data on in situ concentrations of biologically important gases (carbon dioxide and oxygen), coupled with enumeration of total and specific degrader populations of microorganisms, may provide sufficient data to allow a decision to be made to implement closure based on natural attenuation monitoring.

CONCLUSION

As with any scientific investigation, one can never have enough information. However, what the above scenario has illustrated is the difficulty in trying to extrapolate backwards to determine whether or not conditions are, or continue to be, conducive for microbiological activity. However, this does not preclude the use of natural attenuation for site closure if a proper monitoring program is implemented. It is very beneficial to consider the potential implementation of bioremediation at the beginning of any evaluation of remedial strategy for PHC groundwater and soil remediation. Many of the parameters that are useful for tracking bioremediation progress can easily be implemented during the required site-assessment groundwater-monitoring activities. At a minimum, these parameters include temperature, pH, dissolved oxygen or alternative electron acceptors, redox potential, and microbial enumerations.

ACKNOWLEDGMENTS

The authors would like to acknowledge the support of Vin Maresco and Annmarie Kearns of GES-NY.

REFERENCES

Baker, K.H., and D.S. Herson. 1994. *Bioremediation*. McGraw-Hill Book Co. New York, NY.
Borden, R.C. 1993. "Natural Bioremediation of Hydrocarbon-Contaminated Ground Water." In *In-Situ Bioremediation of Groundwater and Geological Materials: A Review of Technologies*. U.S. Environmental Protection Agency, EPA 600/R-93/124, Robert S. Kerr Environmental Research Laboratory, Ada, OK.
Gerhardt, P., R.G. Murray, W. A. Wood, and N.R. Krieg (Eds.). 1994. *Methods for General & Molecular Bacteriology*. American Society for Microbiology, Washington, DC.
Mercer, J.W., D.C. Skipp, D. Giffin, and R.R. Ross. 1990. *Basics of Pump-and-Treat Groundwater Remediation Technology*. U.S. Environmental Protection Agency, EPA-600/6-90/003, Robert S. Kerr Environmental Research Laboratory, Ada, OK.
National Research Council. 1993. *In Situ Bioremediation: When Does It Work?* National Academy Press, Washington, DC.

Intrinsic Bioremediation of Jet Fuel Contamination at George Air Force Base

John T. Wilson, Guy Sewell,
Denise Caron, Greg Doyle,
and Ross N. Miller

ABSTRACT

The rate of intrinsic bioremediation of BTEX compounds in groundwater from a spill of JP-4 jet fuel was estimated by comparing attenuation of the concentrations of the compounds along a flow path. Concentrations of the trimethylbenzenes (TMB) were used to correct for attenuation due to dilution. Analysis of core samples identified the depth interval in the aquifer that was occupied by the groundwater plume. A downhole flowmeter test identified the local hydraulic conductivity of the depth interval occupied by the plume. Time of travel between wells along the flowpath was calculated from the hydraulic gradient and hydraulic conductivity, assuming an effective porosity of 0.3. First-order rate constants were calculated from attenuation (corrected for dilution or dispersion) and the estimated residence time of groundwater between the wells.

INTRODUCTION

Groundwater at George Air Force Base (AFB), California has been impacted by one or several releases of JP-4 jet fuel. Water from monitoring wells on the site was examined for evidence of natural attenuation. Most of the wells indicated an attenuation of benzene, toluene, ethylbenzene, and xylenes (BTEX) as the groundwater moved away from the fuel spill. The wells were examined for the depletion of natural electron acceptors, and for the presence of fuel components that could be used as tracers to correct for dilution.

Figure 1 presents the location of the wells, the contour of the water table, and the boundary between areas with free product in wells and the area without free product. Well MW-24 had 2.06 m of floating jet fuel on 3/25/92, and 2.11 m on 4/21/92.

EXPERIMENTAL PROCEDURES

The monitoring wells were 10-cm internal diameter (I.D.) polyvinyl chloride wells, screened from the water table to a depth 9.1 m below the water table. The water table was 40 m below the land surface. Water samples were preserved with trisodium phosphate and analyzed by headspace analysis using gas chromatography (GC)/mass spectrometry (MS). Core samples were acquired through a 10-cm I.D. hollow-stem auger with a split-spoon core barrel (46 cm long) on the end of a wire line hammer. Core samples were subsampled in the field (approximately 15 g per subsample) and extracted with 5 mL of methylene chloride. The extracts were analyzed in the laboratory by GC/MS.

Flow monitoring was conducted using a prototype downhole flowmeter developed for the U.S. Environmental Protection Agency (EPA) by the Tennessee Valley Authority. Flow was measured at various depth intervals in wells while the wells were being pumped from near the water surface at a rate of 20 L/min. The flow through the flowmeter at any one depth included all the water collected by the well below that depth interval. Local hydraulic conductivity was estimated by comparing the net contribution to flow between adjacent vertical intervals.

RESULTS

Use of Trimethylbenzenes to Correct for Dilution

Table 1 presents data from a flow path on the eastern margin of the plume downgradient of the JP-4 spill. Well MW-24 is near the center of the spill, and MW-25 is near the edge of the spill, 150 m from MW-24 in a direction perpendicular to groundwater flow (Figure 1).

Water from MW-24 is typical of the water from the center of the spill. Water from MW-25 contains both electron acceptors and hydrocarbons, indicating that it samples the spill as well as unimpacted water beneath the spill. Wells MW-27, MW-29, and MW-31 are along a flow path downgradient of MW-25. Oxygen, nitrate, and sulfate were depleted downgradient of the spill.

In MW-24, MW-25, MW-27, and MW-29, the ratios of the concentrations of the three trimethylbenzenes (TMBs) were conserved (Table 1), although the absolute concentrations of the TMBs changed several fold. This suggests that the changes were due to simple dilution and not biodegradation. Compare MW-24 (representative of the core of the spill) and MW-29 in the plume (Table 1). The concentration of benzene was reduced more than 300-fold in MW-29, but the concentration of 1,2,3-TMB was only reduced 3-fold. After correcting for dilution, benzene concentrations were reduced at least 100-fold along this transect because of biological activity.

A second flow path from near the center of the spill followed a different pattern. The TMBs were attenuated along with benzene, toluene, ethylbenzene,

TABLE 1. Relative attenuation of BTEX compounds and TMBs along a flow-path undergoing natural attenuation.

Element or compound	MW-24 Center of oil lens	MW-25 Edge of oil lens	MW-27 210 m away	MW-29 360 m away	MW-31 550 m away
	(mg/liter)				
Oxygen	<0.5	8.0	0.6	<0.5	1.1
Nitrate-N	0.8	3.7	0.4	0.3	3.1
Sulfate	99	170	127	173	225
Alkylbenzenes	4.7	1.5	0.39	0.083	< 0.001
Compound	(μg/liter)				
Benzene	1620	194	80	4.8	<0.5
Toluene	1500	604	<0.5	<0.5	<0.5
Ethylbenzene	210	92	117	2.1	<0.5
p-Xylene	182	83	2.8	2.6	<0.5
m-Xylene	522	240	<0.5	<0.5	<0.5
o-Xylene	377	144	7.3	<0.5	<0.5
1,3,5-TMB [a]	38	22	27	9.3	<0.5
1,2,4-TMB	106	57	69	28	0.8
1,2,3-TMB	73	39	56	20	<0.5

(a) TMB is trimethylbenzene.

and xylenes (BTEX) (Table 2). Benzene was relatively recalcitrant. The only indicator that attenuation was not due to dilution is the absence of nitrate and oxygen in the downgradient wells.

Depletion of Electron Acceptors

Comparing MW-25 and MW-27 in Table 1, 7.4 mg/L oxygen, 3.3 mg/L nitrate-N, and 4.3 mg/L sulfate were consumed. Following the stoichiometry of Wilson et al. (1994), the depletion of oxygen, nitrate, and sulfate would account for the removal of 2.4, 3.1, and 9.0 mg/L of aromatic petroleum hydrocarbons.

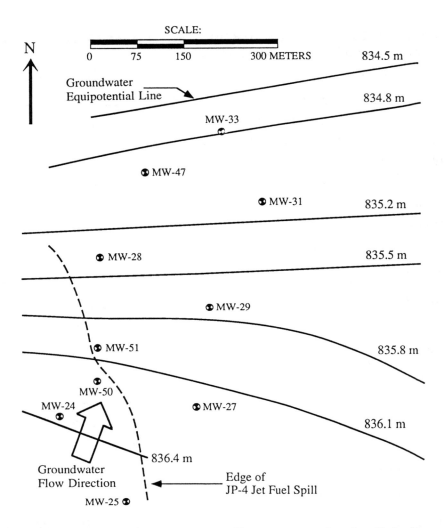

FIGURE 1. Location of monitoring wells near or under the flight line at George AFB, CA. The arrow indicates the direction of groundwater flow. The water table is contoured with an interval of 1 ft (0.305 m).

Based on concentrations in MW-25 and MW-27, approximately 1.1 mg/L of total alkylbenzenes, including TMBs and tetramethylbenzenes, was removed from the groundwater. The electron acceptor demand exerted in the flow path was sufficient to account for the observed removal of petroleum hydrocarbons. Perhaps additional aromatic hydrocarbons were recruited from the "floating" jet fuel as groundwater moved from MW-25 to MW-27.

In both flow paths, the concentration of sulfate increased with distance away from the spill. This would be expected if sulfate were depleted during

TABLE 2. A flowpath showing attenuation of TMBs and BTEX compounds.

	MW-24	MW-50	MW-51	MW-28	MW-47
	Center of oil lens	Edge of oil lens	60 m away	150 m away	300 m away
Element or compound			(mg/L)		
Oxygen	<0.5	<0.5	<0.5	<0.5	1.1
Nitrate-N	0.8	<0.05	<0.5	0.3	1.6
Sulfate	89	79	92	111	190
Compound			(µg/L)		
Benzene	1,620	859	4.9	0.7	<0.5
Toluene	1,500	1,510	2.6	<0.5	<0.5
Ethylbenzene	210	159	<0.5	1.2	<0.5
p-Xylene	182	134	<0.5	<0.5	<0.5
m-Xylene	522	378	1.0	<0.5	<0.5
o-Xylene	377	254	1.0	<0.5	<0.5
1,3,5-TMB[a]	38	28	4.2	<0.5	<0.5
1,2,4-TMB	106	78	<0.5	<0.5	<0.5
1,2,3-TMB	73	50	8.3	3.4	<0

(a) TMB is trimethylbenzene.

transit of groundwater under the spill, then was replenished by admixture with uncontaminated water as the plume moved away from the spill, or by dissolution of sulfate minerals from the aquifer matrix.

Comparison of Water Quality Predicted from Core Analysis with Quality of Water Produced by a Monitoring Well

The concentrations of alkylbenzenes in core samples from the aquifer were used to predict the concentration in groundwater from that interval. Compound concentration in μg/kg sediment was divided by an assumed bulk density of 1.9 to estimate μg/cubic decimeter of aquifer material, then divided by an assumed porosity of 0.3 to estimate μg/L groundwater. The analytical detection limit was equivalent to 130 μg/L.

Figure 2 presents the predicted concentration of benzene in groundwater for all the core samples that had benzene present above the detection limit.

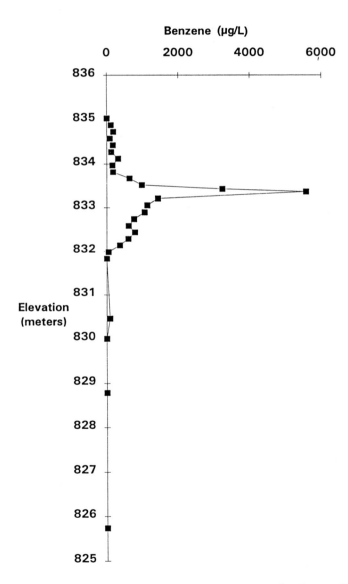

FIGURE 2. Vertical distribution of benzene in core samples from a JP-4 release near MW-24.

Appreciable concentrations of benzene were confined to a depth interval with an elevation from 834 to 832 m. Calculations were performed for all the alkyl-benzenes for two core samples collected at elevations of 833.3 m and 832.4 m. Table 3 compares the predicted concentrations to concentrations in well water on two separate occasions.

Concentrations of alkylbenzenes in the water from MW-24 were reasonably consistent between the two sampling events, particularly for the TMBs. The highest concentration of benzene predicted from the core samples was only twice the concentration in the well water, and the average over the interval from 40.5 m to 42.1 m was very similar to the concentration of water produced by MW-24.

Results of the Downhole Flowmeter Testing

The following data are the personal communication of Randall Ross and Steve Acree (R. S. Kerr Lab, U.S. EPA). Figure 3 presents the vertical

TABLE 3. Comparison of groundwater produced from monitoring wells to groundwater quality predicted from core analysis.

	Monitoring Well		Predicted from Cores	
	MW-24 1/5/94	MW-24 4/26/94	Near MW-24 40.9 m deep	Near MW-24 average 40.5 to 42.1 m
Compound		(μg)		
Benzene	1620	1400	5500	1236
Toluene	1520	541	1800	836
Ethylbenzene	210	163	378	
p-Xylene	182	141	485	
m-Xylene	522	241	1040	630
o-Xylene	377	340	584	503
1,3,5-TMB[a]	37.8	36.2	89	
1,2,4-TMB	106	97.6	288	
1,2,3-TMB	72.9	65.8	173	<130

(a)TMB is trimethylbenzene.

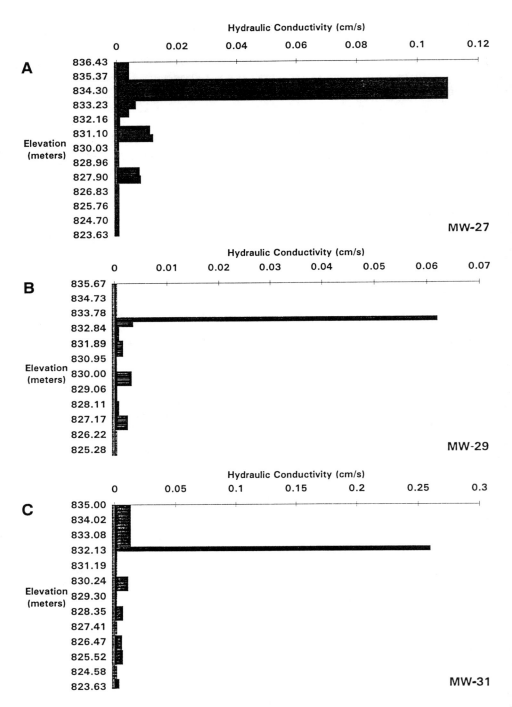

FIGURE 3A,B,C. Vertical distribution of hydraulic conductivity in the aquifer sampled by wells MW-27, MW-29, and MW-31.

distribution of hydraulic conductivity in monitoring wells MW-27, MW-29, and MW-31, which are arranged along a flowpath along the east side of the ground-water plume. Although 9 vertical meters of aquifer was sampled by the wells, most of the flow was contributed by a narrow interval from 833 m to 835 m in elevation. As the plume moves north from the spill, the conductive depth interval in the aquifer is thinner, but more conductive. The depth interval revealed by flowmeter testing to have the maximum hydraulic conductivity is equivalent to the depth interval containing the groundwater plume as revealed by analysis of core samples (compare Figure 2).

The flowmeter is not intrinsically safe in the presence of fuel. Well MW-25 has 0.2 m of floating fuel, and no estimate of local hydraulic conductivity is available. To estimate plume residence time between MW-25 and MW-27, we assumed that the maximum local hydraulic conductivity at MW-27 (0.11 cm/s) applied to the entire interval. The hydraulic gradient was 0.004 at the time the flowmeter testing was done. Hydraulic conductivity multiplied by hydraulic gradient is darcy flow. We divided the darcy flow by an assumed effective porosity of 0.3 to predict an interstitial seepage velocity of 8.87 m/week. To estimate plume residence time between MW-25 and MW-29, we averaged the maximum local hydraulic conductivity at MW-29 (0.062 cm/s) with the local hydraulic conductivity at MW-27. The predicted seepage velocity was 6.83 m/week.

The residence time is simply the distance between the wells along the flow path, divided by seepage velocity. Based on conventional aquifer tests, on the hydraulic gradient, and using the assumption that flow in the aquifer is uniform, the estimated interstitial seepage velocity of the plume should be 0.12 m/week. This "conventional" approach would overestimate residence time by more than an order of magnitude.

Following the method of Wiedemeier et al. (1995a), the concentrations of 1,2,4-TMB were used to correct for dilution. In Table 4 the calculated rate con-stants are compared to calculated rates from similar releases at other sites (Hill AFB, Utah; Eglin AFB, Florida) with similar groundwater temperatures (Wiede-meier et al. 1995b; Wilson et al. 1994). Compared to the other two sites, rates of benzene biodegradation at George AFB are slow. Both George AFB and Hill AFB are in arid upland landscapes, and sulfate reduction is the dominant sink for fuel hydrocarbons at both locations. Eglin AFB is in a weathered coastal environment, and the dominant sink is methanogenesis.

No information is available at George AFB on the vertical distribution of sulfate and other electron acceptors. These electron acceptors may be depleted in the local cross section of aquifer that is actually carrying the plume. If such is the case, then mass transport processes (such as diffusion of electron accep-tors or admixture of groundwater that has not been depleted of electron accep-tors) will control the rate of degradation, not the metabolic capability of the microorganisms.

TABLE 4. Comparison of provisional first-order rates of degradation along a flowpath at George AFB to rates reported for plumes of petroleum hydro-carbons from JP-4 releases at other sites.

Compound	MW-25 to MW-27 23.7 weeks travel	MW-25 to MW-29 51.9 weeks travel	Hill AFB, UT	Eglin AFB, FL
	First-Order Rate Constant (per week)			
Benzene	0.053	0.058	0.23 and 0.62	none
Toluene	>0.32		0.74	none
Ethylbenzene	none	0.060	0.17 and 0.56	0.21 and 0.38
p-Xylene	0.16		0.13 and 0.22	0.14 and 0.57
m-Xylene	>0.28		0.18	0.14 and 0.73
o-Xylene	0.14		0.16	0.015

DISCLAIMER

Although the research described in this article has been funded wholly or in part by the U.S. Environmental Protection Agency, it has not been subjected to U.S. EPA review, and therefore does not necessarily reflect the views of the agency, and no official endorsement should be inferred.

REFERENCES

Wiedemeier, T. H., D. C. Downey, J. T. Wilson, D. H. Kampbell, R. N. Miller, and J. E. Hansen. 1995a. "Technical Protocol for Implementing Intrinsic Remediation with Long-Term Monitoring for Natural Attenuation of Fuel Contamination Dissolved in Groundwater." (Draft)

Wiedemeier, T. H., M. A. Swanson, J. T. Wilson, D. H. Kampbell, R. N. Miller, and J. E. Hansen. 1995b. "Patterns of Intrinsic Bioremediation at Two United States Air Force Bases." In R. E. Hinchee, J.T . Wilson, and D. C. Downey (Eds.), *Intrinsic Bioremediation*, pp. 31-51. Battelle Press, Columbus, OH.

Wilson, J. T., F. M. Pfeffer, J. W. Weaver, D. H. Kampbell, T. H. Wiedemeier, J. E. Hansen and R. N. Miller. 1994. "Intrinsic Bioremediation of JP-4 Jet Fuel." *Symposium on Intrinsic Bioremediation of Groundwater*. U.S. Environmental Protection Agency Report, EPA/540/R-94/515.

Anaerobic BTEX Biodegradation in Laboratory Microcosms and In Situ Columns

Melody J. Hunt, Michael A. Beckman,
Morton A. Barlaz, and Robert C. Borden

ABSTRACT

Assessing the potential for natural bioremediation in the subsurface is complicated by site-specific conditions and the methods used to estimate biodegradation rates. Controlled laboratory experiments often are necessary to verify biological loss of a compound and to assess factors that influence biodegradation. However, the effect of removing samples from such a stable environment and placing them in laboratory microcosms is not fully understood. Sample removal clearly changes spatial relationships in the indigenous microbial community and could influence results obtained in the laboratory. In situ columns have been used to measure biodegradation on a limited basis, and little is known about their reliability. The objective of this paper is to evaluate the use of laboratory microcosms and in situ column experiments for estimating intrinsic biodegradation rates of BTEX (benzene, toluene, ethylbenzene, and xylene isomers) in the subsurface.

SITE BACKGROUND

This research was conducted at a petroleum-contaminated aquifer in the southeastern coastal plain of North Carolina. The plume is characterized by negligible dissolved oxygen and redox potentials of -100 to -200 mV due to intrinsic biodegradation of BTEX (Borden et al. 1995). The dominant electron acceptors within the plume are sulfate and iron (Borden et al. 1994). The midpoint of the plume is characterized by high dissolved Fe^{2+} (>40 mg/L) and low SO_4^{2-} concentrations (<4 mg/L). Toluene and o-xylene are nearly depleted (<20 μg/L), whereas high quantities of benzene, ethylbenzene, and m-, and p-xylenes remain (> 500 μg/L).

MATERIALS AND METHODS

In situ column and laboratory microcosm experiments were conducted to measure the indigenous microorganisms' ability to anaerobically degrade BTEX. Both measurements consist of spiking aquifer sediment with BTEX and monitoring compound disappearance over time. Killed controls were monitored at the same time to differentiate between biological and abiotic losses. Analytical methods are described elsewhere (Beckman 1994, Hunt et al. 1995).

In Situ Columns

The in situ columns (Figure 1) are similar to the system used by Gillham et al. (1990). Each column consists of a 1-m-long chamber where sediment and groundwater are isolated from the surrounding aquifer for controlled observation. Columns were installed by drilling a pilot hole and then installing a 15-cm-diameter by 3-m-long section of polyvinyl chloride (PVC) casing.

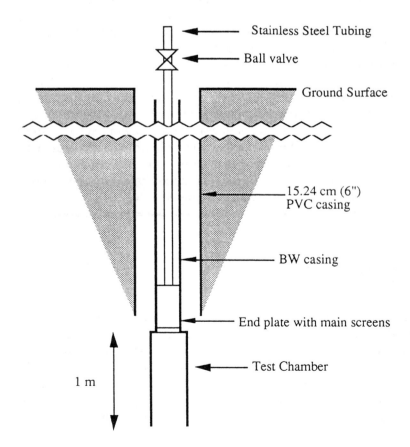

FIGURE 1. In situ column schematic.

Stainless steel tubing and 3 m of drill rod were attached to the equipment chamber. Argon was pumped into the casing to displace any oxygen present. The column was then pushed into the aquifer while applying suction to the stainless steel feed line to ensure that the chamber completely filled with aquifer material. The columns were then filled with anaerobic groundwater containing BTEX, which had been recovered from a nearby well.

Two sets of columns, Group A and Group B, were installed at the midpoint area of the plume. Each set contained three individual columns; two live and one abiotic control. The abiotic control columns were prepared by adjusting the pH to <2 with HCl. Tracer tests were conducted on all columns prior to the start of each experiment to ensure that they were properly installed. Iron and dissolved oxygen (DO) remained constant during the tracer test, indicating that oxygen was not introduced. Possible sorption losses for each compound were then estimated using BTEX concentrations observed in an adjacent multilevel well screened at the same interval as the column (Beckman 1994).

All columns were monitored on a monthly basis for BTEX, dissolved iron, sulfate, chloride, pH, and DO. Group A columns were monitored for 7 months, while Group B columns were monitored for 3 months.

Laboratory Microcosms

Multiple replicate microcosms with no headspace were constructed in the laboratory under aseptic conditions using blended aquifer sediment and groundwater recovered under anaerobic conditions. Microcosm preparation was designed to simulate ambient conditions to the maximum extent possible and is described in more detail elsewhere (Hunt et al. 1994). Briefly, microcosms were spiked with approximately 10,000 μg/L BTEX (2,000 μg/L of each compound) and incubated in anaerobic containers stored at the ambient groundwater temperature, 16°C. The compound *p*-xylene was not added to the microcosms since it cannot be distinguished from *m*-xylene by the analytical procedure used. BTEX loss was monitored by destructively sampling three live and three abiotic microcosms at monthly intervals for 300 days. A final timepoint was taken 100 days later (after 400 days of incubation).

IN SITU COLUMN RESULTS

All live and abiotic columns exhibited an initial concentration decrease of several hundred μg/L between the injection water and the first sample taken from the chamber. This initial loss was attributed to sorption to the aquifer sediment (Beckman 1994). After the sorption loss, the initial compound concentrations were less than 500 μg/L in most columns. The concentrations of hydrocarbons in the abiotic columns remained fairly constant or declined slowly after the initial drop, indicating biological activity or short circuiting did not occur in the control columns. The dissolved oxygen concentration in the control columns remained low (<0.4 mg/L) throughout the entire experiment.

In Group A, benzene and *m-, p*-xylene exhibited significantly higher losses in the live columns relative to the abiotic columns. The concentration of *m-, p*-xylene decreased in the live columns after an initial lag of between 85 and 121 days. Benzene concentrations remained constant in both live columns for a period of approximately 155 days, after which time decreases attributed to biological activity were measured. Initial toluene and *o*-xylene concentrations were too low (<50 μg/L) to measure concentration changes accurately. In Group B, there was significant biological loss of toluene with no apparent lag time. The short sampling period of 75 days was not adequate to measure losses of the other compounds in the Group B columns.

Effective first-order removal rates for BTEX were estimated using the equation $C = C_0 \exp(-Kt)$, where K is the apparent first-order decay rate (d^{-1}), t is time, and C_0 is the initial concentration (Table 1). Although first-order rates are reported, it should be noted that the measured compound losses could also be

TABLE 1. Apparent first-order decay rates for laboratory microcosms and in situ columns.

	Time Period[a] (days)	Live Decay Rate[b] (day^{-1})	Abiotic Decay Rate [c] (day^{-1})	Biological Loss[d] (day^{-1})	Degrees Freedom	Signifi-cance[e]
In Situ Columns						
Benzene	155 - 251	0.0060	0.0010	0.0049	12	>99%
Toluene	13 - 75	0.0169	0.0054	0.0115	11	>99%
Ethylbenzene	15 - 251	0.0019	0.0008	0.0011	22	N.S.[f]
o-Xylene	13 - 123	0.0021	0.0040	−0.0019	17	N.S.
m-, p-Xylene	121 - 251	0.0194	0.0051	0.0143	12	>99%
Laboratory Microcosms						
Benzene	184 - 403	0.0258	0.0021	0.0237	19	>99%
Toluene	22 - 120	0.0489	0.0043	0.0446	20	>99%
Ethylbenzene	0 - 403	0.0056	0.0037	0.0019	50	>95%
o-Xylene	37 - 120	0.0611	0.0052	0.0559	14	>99%
m-Xylene	0 - 184	0.0234	0.0030	0.0204	37	>99%

(a) Time period of measured decay.
(b) Measured loss in live samples for time period indicated.
(c) Measured loss in killed controls (abiotic) samples for time period indicated.
(d) Difference between live and abiotic loss.
(e) Confidence level that difference exists between measured live and abiotic decay rates determined by students *t* tests assuming unequal variances.
(f) Not significant.

fitted to a zero-order decay model. Decay rates were calculated over the time period in which biological losses were observed. The rate of biological loss is represented as the difference between the decay rates in the live and abiotic columns taken over the same time period. Figure 2 shows the measured *m-*, *p-*xylene loss in the Group A columns, and illustrates the time frame used to calculate the decay rates in the live and abiotic columns. Results from the two live columns were pooled to estimate the live decay rate.

LABORATORY MICROCOSM RESULTS

A distinct order of compound disappearance was measured in the laboratory incubations: *m-*xylene degradation began with no lag period, followed by toluene, *o-*xylene, benzene, and ethylbenzene (Figure 3). The rate of *m-*xylene loss declined once toluene loss began (after 22 days), and did not increase until the toluene and *o-*xylene were below 20 μg/L (120 days). The decay rate of *m-*xylene varied from 0.052 day^{-1} initially (days 1 to 37) to 0.011 day^{-1} (days 37 to 120), while toluene and *o-*xylene losses were measured. The aquifer material was obtained in an area of the plume where toluene and *o-*xylene concentrations were very low (<50 μg/L) but significant quantities of *m-*, *p-*xylene remained (>1,000 μg/L). Thus, the microbial population appeared to have an initial preference for *m-*xylene, but switched to toluene and *o-*xylene after a 22-day acclimation period. Benzene began to biodegrade once *m-*xylene was depleted and was at or below 10 μg/L in all microcosms at the final sampling (403 days). As with the column experiments, first-order decay rates (K) were

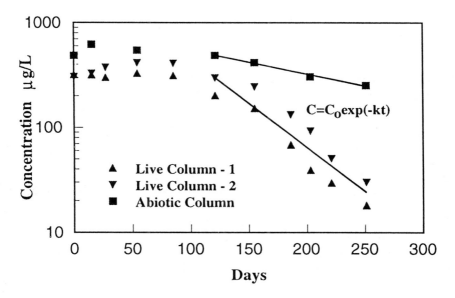

FIGURE 2. Anaerobic decay of *m-*, *p-*xylene in group A in situ columns.

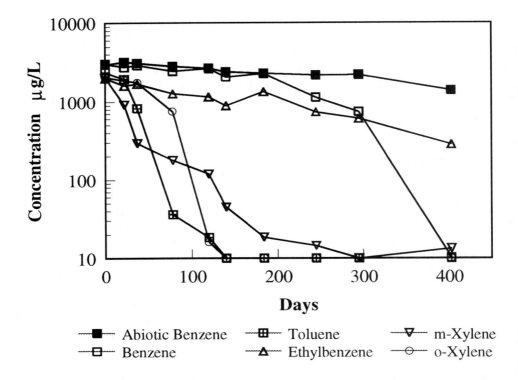

FIGURE 3. Anaerobic BTEX loss in laboratory microcosms.

determined during the time period following measured loss for each compound (Table 1).

COMPARISON OF IN SITU COLUMNS AND LABORATORY MICROCOSMS

Biological loss for three of the five BTEX compounds occurred over similar time periods in the laboratory and in situ experiments. In both cases, toluene degradation was most rapid, followed by *m-*, *p*-xylene and benzene. This order is consistent with previous field investigations (Borden et al. 1995). Significant concentration decreases did not occur concurrently for these compounds in column or microcosm experiments. Ethylbenzene loss was minimal in the laboratory microcosms during the 400 days of incubation, and was not detected during the 7 months of in situ monitoring. Loss of *o*-xylene was not observed in the Group B columns, but fairly rapid depletion concurrent with toluene loss was measured in the laboratory. It is possible that the initial concentration of *o*-xylene (<500 µg/L) was too low to stimulate in situ degradation or that the 75-day monitoring period was too short.

Although the monthly sampling frequency was consistent for both types of measurements, the time of monitoring was shorter in the in situ columns because of the limited sample volume available. Thus, direct comparison of decay rates between the two types of measurements is difficult. More frequent sampling would be needed to accurately measure decay rates for some compounds in both types of experiment. Given these limitations, the measured rates are comparable in both columns and microcosms for most of the compounds, but are an order of magnitude higher than rates estimated from field investigations (Borden et al. 1994). Both column and laboratory measurements were conducted at one point (mid-plume area), whereas the field estimates were made along the length of the plume centerline. Microcosm studies conducted with material from other areas of the plume (Hunt et al. 1994) show severe variation in BTEX decay. Thus, one explanation for the lower field rates is that the field estimate is an average along the plume length, while the microcosm and column experiments are indicative of decay at one point within the plume.

Biological decay was demonstrated in the controlled column and microcosm experiments. Use of in situ columns could provide a practical link between laboratory evaluations and full-scale field studies.

ACKNOWLEDGMENTS

The research described in this article was supported in part by the Robert S. Kerr Laboratory of the U.S. Environmental Protection Agency under a cooperative agreement, CR 819630-01-0; the National Ground Water Association; and the Hoechst Celanese Corporation. We would also like to thank the property owner, Mr. H. Richardson, for access to the site.

REFERENCES

Beckman, M. A. 1994. "In-Situ Measurement of Intrinsic Bioremediation Rates." Master of Science Thesis, Department of Civil Engineering, North Carolina State University, Raleigh, NC.

Borden, R. C., C. A. Gomez, and M. T. Becker. 1994. "Natural Bioremediation of a Gasoline Spill." In R. E Hinchee, B. C. Alleman, R. E. Hoeppel, and R. N. Miller (Eds.), *Hydrocarbon Bioremediation.* pp. 290-295. Lewis Publishers, Boca Raton, FL.

Borden, R. C., C. A. Gomez, and M. T. Becker. 1995. "Geochemical Indicators of Natural Bioremediation." *Ground Water 33*: 180-189.

Gillham, R. W., R. C. Starr, and D. J. Miller. 1990. "A Device for In-Situ Determination of Geochemical Transport Parameters; 2 Biochemical Reactions." *Ground Water 82*: 858-862.

Hunt, M. J., Borden, R. C., and M. A. Barlaz. 1995. "Rate and Extent of Natural Anaerobic Biodegradation of BTEX Compounds in Groundwater Plumes," 87th Annual Meeting of the Air & Solid Waste Management Association, Cincinnati, OH, June 19-24 (in press).

Regression Techniques and Analytical Solutions to Demonstrate Intrinsic Bioremediation

Timothy E. Buscheck and Celia M. Alcantar

ABSTRACT ━━━━━━━━━━━━━━━━━━━━━━━━━━━━━━━━━━━

It is now generally recognized that a major factor responsible for the attenuation and mass reduction of benzene, toluene, ethylbenzene, and xylenes (BTEX) in groundwater plumes is hydrocarbon biodegradation by indigenous microorganisms in aquifer material. Our objective is to apply well-known regression techniques and analytical solutions to estimate the contribution of advection, dispersion, sorption, and biodecay to the overall attenuation of petroleum hydrocarbons. These calculations yield an apparent biodecay rate based on field data. This biodecay rate is a significant portion of the overall attenuation in stable, dissolved hydrocarbon plumes.

INTRODUCTION

"Intrinsic bioremediation" is the degradation of organic compounds by indigenous microbes without artificial enhancement. Advection, dispersion, sorption, and decay each contribute to the overall attenuation of a dissolved hydrocarbon plume. The effect of advection is to transport dissolved contaminants at the same rate as the groundwater velocity. The effect of dispersion is to spread contaminant mass beyond the volume it would occupy due to advection alone, and reduce contaminant concentrations. The effect of sorption is to retard contaminant migration. These factors affect the configuration of dissolved hydrocarbon plumes. Overall attenuation can cause a plume to shrink over time, create a stable plume, or reduce the rate of plume migration. Two of the conditions for which intrinsic bioremediation is likely to contribute to the configuration of a contaminant plume are a shrinking plume and a stable plume. The configuration of a migrating plume can also be affected by intrinsic bioremediation. Under the conditions of a shrinking plume, degradation mechanisms are necessarily present. Intrinsic bioremediation also is likely to contribute to a stable plume, particularly if the source persists in residually contaminated soils at the water table. In this paper we couple the regression of concentration versus distance for stable plumes to an analytical solution for one-dimensional, steady-state, contaminant transport. The analytical solution includes advection, dispersion, sorption, and decay.

Biological transformation is the process that likely contributes most to the decay of compounds such as BTEX. Several studies suggest the concurrent loss of electron acceptors from groundwater as an indicator of biodegradation (McAllister & Chiang 1994; Salanitro 1993). The mechanism of biodegradation is complex, and the rate is most likely controlled by the mixing of the contaminant and electron acceptors in a three-dimensional, heterogeneous aquifer. The assumption of a first-order decay is a useful approximation of this complex phenomenon. Evaluation of site data suggests apparent first-order attenuation rates occur in the range of 0.1 to 1.0% per day (Buscheck et al. 1993).

The objective of this paper is to provide tools to assist in documenting the loss of contaminants. The regression techniques and analytical solution described are intended to distinguish those mechanisms that contribute to contaminant loss.

PLUME CHARACTERISTICS

Shrinking Plume

Dissolved hydrocarbon plumes may decrease in size, as observed by declining contaminant concentrations in monitoring wells. Exponential regression methods can be used to evaluate whether concentration versus time data fit a first-order decay observed for petroleum hydrocarbons under certain conditions. The solution to the first-order decay is:

$$C(t) = C_i e^{-(kt)} \tag{1}$$

Where $C(t)$ (M/L^3) is concentration as a function of time, t (T), C_i is the initial concentration at $t = 0$, and k is the first-order attenuation rate, T^{-1}. Equation (1) may be used to evaluate contaminant concentration versus time data for individual monitoring wells.

Stable Plume

A stable plume is characterized by dissolved contaminant concentrations remaining constant over time in individual monitoring wells. Short-term variations in monitoring well concentrations due to water table fluctuation, variability in groundwater flow direction, sampling variability, and analytical uncertainty should be distinguished from statistically significant concentration changes. In order for a plume to reach stable conditions, the rate of natural attenuation must be equal to the rate of contaminant addition to the aquifer from the source (McAllister & Chiang 1994). The contaminant source or influx rate is limited by the compound's effective solubility and the flow rate of water through the source area (infiltration, fluctuating water table, etc.).

Kemblowski et al. (1987) recast equation (1) for concentration as a function of distance:

$$C(x) = C_o e^{-\left(k\frac{x}{v_x}\right)} \tag{2}$$

Where C_o (M/L^3) is the concentration at the source. The transformation of the exponential terms in equations (1) and (2) is achieved by substituting time, t, with distance traveled, $x(L)$ divided by the linear groundwater velocity, v_x (L/T). The term "x/v_x" is the residence time for pore water to move some distance, x, from the source. The concentration versus distance regression is based on equation (2). The groundwater flow direction is defined based on multiple monitoring events covering the hydrologic cycle. Six monitoring wells were selected along the groundwater flow path (see inset of Figure 1). A minimum of three monitoring wells are required for this analysis. In this case, contaminant concentrations declined with downgradient distance. Figure 1 plots benzene concentration versus distance for a terminal in Fairfax, Virginia. From the exponent of equation (2), the slope of the line in Figure 1 is k/v_x (L^{-1}), the reciprocal of the attenuation distance. If this slope is multiplied by groundwater velocity (L/T), we obtain k (T^{-1}). In the absence of a reliable estimate of groundwater velocity, the k/v_x term is useful, particularly for estimating the downgradient extent of contaminant migration and selecting downgradient monitoring well locations.

ANALYTICAL SOLUTION FOR A STABLE PLUME

The general one-dimensional transport equation, with first-order decay of the contaminant, is given by the following equation:

$$\frac{\delta C}{\delta t} = \frac{1}{R_f}\left[D_x \frac{\delta^2 C}{\delta x^2} - v_x \frac{\delta C}{\delta x}\right] - \lambda C \tag{3}$$

Where D_x (L^2/T) is the dispersion coefficient, v_x (L/T) is the linear groundwater velocity, R_f $(-)$ is the retardation coefficient, and λ (T^{-1}) is the total decay rate. The form of equation (3) assumes D_x is constant and independent of distance, x. While the terms in brackets describe the mass transport by dispersion and advection, respectively, the retardation coefficient characterizes the contribution of sorption. The form of this equation assumes degradation occurs in the aqueous and sorbed phases at the same rate. If biological transformation of BTEX compounds occurs primarily in the aqueous phase, the term "λC" would appear inside the brackets.

Dispersion and advection are related by the longitudinal dispersivity, α_x (L), which has been described by empirical expressions (Fetter 1993).

$$D_x = \alpha_x v_x \tag{4}$$

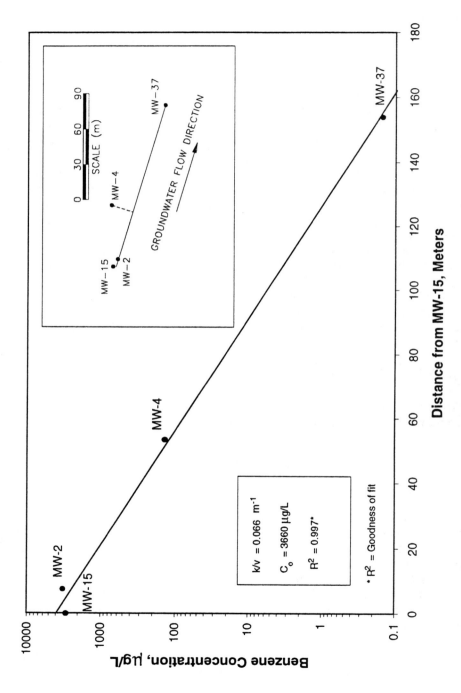

FIGURE 1. Exponential regression of concentration versus distance for Fairfax Terminal.

The retardation coefficient (R_f) accounts for contaminant partitioning between the solid and aqueous phases. R_f describes the relationship between the linear groundwater velocity, and contaminant velocity, v_c (L/T):

$$R_f = \frac{v_x}{v_c} \tag{5}$$

Chiang et al. (1989) demonstrated that the contribution of volatilization to the dissolved contaminant attenuation was only 5% at one site. Except in the case of very shallow groundwater, volatilization is not expected to contribute significantly to the overall attenuation. Therefore, volatilization is neglected and the decay rate is assumed to be a measure of biodegradation of BTEX compounds.

Bear (1979) solved equation (3) for concentration. The steady-state solution is given as:

$$C(x) = C_o \exp\left[\left(\frac{x}{2\alpha_x}\right)\left[1-\left(1+\frac{4\lambda\alpha_x}{v_c}\right)^{1/2}\right]\right] \tag{6}$$

For the case in which decay occurs only in the aqueous phase, the contaminant velocity, v_c, is replaced by the linear groundwater velocity, v_x, in equation (6). As the decay rate (λ) increases with respect to the other transport mechanisms, the concentration away from the source ($x > 0$), approaches zero because the material is decaying at a greater rate than it is being transported through the medium. Similarly, as the contaminant velocity increases, the decay becomes less effective in reducing concentrations as a function of distance. Retarded contaminants therefore have a greater opportunity to decay because retarded transport velocities favor biodegradation kinetics over transport (Domenico & Schwartz 1990).

The exponential regression for concentration versus distance yields the reciprocal of the attenuation distance, k/v_x (L^{-1}), previously shown in equation (2). Equations (2) and (6) are of the same form:

$$C(x) = C_o \exp (mx) \tag{7}$$

The slope of the log-linear data is given by m. The one-dimensional, steady-state transport solution also describes the slope, m, of the log-linear data:

$$m = \left(\frac{1}{2\alpha_x}\right)\left[1-\left(1+\frac{4\lambda\alpha_x}{v_c}\right)^{1/2}\right] \tag{8}$$

Therefore, the term k/v_x and equation (8) both describe the slope of the log-linear data and can be equated to solve for the total decay rate, λ, a measure of intrinsic bioremediation. Dispersivity (α), contaminant velocity (v_c), and k/v_x are input to the following equation to calculate the decay rate.

$$\lambda = \left(\frac{v_c}{4\alpha_x}\right)\left(\left[1 + 2\alpha_x\left(\frac{k}{v_x}\right)\right]^2 - 1\right) \tag{9}$$

For the case in which decay occurs only in the aqueous phase, v_c is replaced by v_x in equation (9).

RESULTS

The results of equating the spatial regression with the steady-state analytical solution for the Fairfax Terminal are presented in Table 1. The values for source concentration (C_o) and k/v_x were regressed using the data plotted in Figure 1. Table 1 includes k and the ratio, λ/k, the contribution of biodecay to the overall attenuation rate (expressed as %). In Case 1, groundwater velocity was 0.06 m/day, based on aquifer pump tests. Retardation was estimated as 2 and dispersivity was estimated as 7.5 m, approximately 5% of the flow field (distance separating the two furthest wells). In Case 1, $\lambda = 0.30\%$/day (0.0030 days^{-1}); λ is 75% of k for this case. The next four cases were performed to evaluate the sensitivity of changing various input parameters. In each of these cases C_o and k/v_x remain constant. In Case 2, the groundwater velocity is reduced by a factor of two (v = 0.03 m/day), which reduces the decay rate by the same factor ($\lambda = 0.15\%$/day). In this case, only half the decay rate is required to maintain the Case 1 concentration decline with distance; as in Case 1, λ is 75% of k in Case 2. In Case 3, the dispersivity is increased by a factor of two ($\alpha = 15$ m) and $\lambda = 0.40\%$/day. More decay is required with a larger dispersivity because more spreading of the contaminant occurs in the direction of groundwater flow; λ is equivalent to k in Case 3.

Cases 4 and 5 were performed to calculate λ assuming biodecay occurs only in the aqueous phase. This is accomplished by replacing v_c with v_x in equation (9) for λ (R = 1 in Table 1 for Cases 4 and 5). Given this revised formulation, the decay rate, λ, is independent of retardation. By limiting decay to the aqueous phase in Case 4, $\lambda = 0.60\%$/day, twice the decay rate in Case 1. In Case 4, λ is 150% of k. Case 5 is similar to Case 4, but dispersivity is reduced to 0.3 m. In Case 5, $\lambda = 0.40\%$/day. Less decay is required with a smaller dispersivity because less spreading of the contaminant plume occurs in the direction of groundwater flow; λ and k are identical in Case 5.

SUMMARY

Contaminant decay is the primary process contributing to a stable configuration of a dissolved contaminant plume. Given a constant source, sorption and dispersion alone are not likely to account for a stable plume. Sorption only retards contaminant velocity, whereas dispersion results in further spreading of the contaminant, reducing concentrations. Decay (biodegradation of BTEX

TABLE 1. Decay rates based on steady-state analytical solution, C_o = 3,660 µg/L, k/v_x = 0.066 m^{-1} (sensitivity on bold input values).

Case	Groundwater Velocity, v_x (m/day)	Retardation Coefficient, R_f	Contaminant Velocity, v_c (m/day)	Dispersivity, α (m)	Attenuation Rate, k (%/day)	Decay Rate, λ (%/day)	λ/k (%)
Case 1	0.06	2	0.03	7.5	0.40	0.30	75
Case 2	**0.03**	2	0.015	7.5	0.20	0.15	75
Case 3	0.06	2	0.03	**15**	0.40	0.40	100
Case 4	0.06	1[1]	0.06	7.5	0.40	0.60	150
Case 5	0.06	1[1]	0.06	**0.3**	0.40	0.40	100

Note: (1) Calculation of λ independent of v_c.

compounds) is the most significant mechanism that accounts for mass loss in a dissolved contaminant plume. The analytical solution for steady-state contaminant transport can be equated to a regression of concentration versus distance (expressed as k/v_x) to solve for the decay rate, λ. The decay rate is a measure of intrinsic bioremediation of petroleum hydrocarbons and can be used in more sophisticated models.

REFERENCES

Bear, J. 1979. *Hydraulics of Groundwater*. McGraw-Hill, New York, NY.

Buscheck, T. E., K. T. O'Reilly, and S. N. Nelson. 1993. "Evaluation of Intrinsic Bioremediation at Field Sites." *Proceedings, Petroleum Hydrocarbons and Organic Chemicals in Ground Water: Prevention, Detection, and Restoration*. National Ground Water Association/API, Houston, TX. pp. 367-381.

Chiang, C. Y., J. P. Salanitro, E. Y. Chai, J. D. Colthart, and C. L. Klein. 1989. "Aerobic Biodegradation of Benzene, Toluene, and Xylene in a Sandy Aquifer — Data Analysis and Computer Modeling." *Ground Water* 27(6): 823-834.

Domenico, P. A., and F. W. Schwartz. 1990. *Physical and Chemical Hydrogeology*. John Wiley & Sons, New York, NY.

Fetter, C. W. 1993. *Contaminant Hydrogeology*. Macmillan Publishing Company, New York, NY.

Kemblowski, M. W., J. P. Salanitro, G. M. Deeley, and C. C. Stanley. 1987. "Fate and Transport of Residual Hydrocarbon in Groundwater — A Case Study." *Proceedings, Petroleum Hydrocarbons and Organic Chemicals in Ground Water: Prevention, Detection, and Restoration*. National Water Well Association/API, Houston, TX. pp. 207-231.

McAllister, P. M., and C. Y. Chiang. 1994. "A Practical Approach to Evaluating Natural Attenuation of Contaminants in Ground Water." *Ground Water Monitoring and Remediation* 14(2): 161-173.

Salanitro, J. P. 1993. "The Role of Bioattenuation in the Management of Aromatic Hydrocarbon Plumes in Aquifers." *Ground Water Monitoring and Remediation* 13(4): 150-161.

Relating BTEX Degradation to the Biogeochemistry of an Anaerobic Aquifer

Simon G. Toze, Terry R. Power, and Gregory B. Davis

ABSTRACT

Trends in chemical and microbiological parameters in a petroleum hydrocarbon plume within anaerobic groundwater have been studied. Previously, microbial degradation of the hydrocarbon compounds had been substantiated by the use of deuterated hydrocarbons to determine natural (intrinsic) degradation rates within the contaminant plume. Here, sulfate concentration decreases, Eh decreases, and hydrogen sulfide and bicarbonate concentration increases are shown to be associated with the contaminant plume. These trends indicate microbial degradation of the benzene, toluene, ethylbenzene, and xylene (BTEX) compounds by sulfate-reducing bacteria. Stoichiometry indicates that other consortia of bacteria play a role in the degradation of the hydrocarbons. Total microbial cell numbers were higher within the plume than in the uncontaminated groundwater. There is, however, no direct correlation between total microbial cell numbers, and BTEX, sulfate, bicarbonate, and hydrogen sulfide concentrations within the plume.

INTRODUCTION

A preferred remediation strategy for BTEX-contaminated groundwater is in situ biodegradation, because it avoids undue disturbance of surface infrastructure and excessive human exposure, as well as being potentially cost effective. Biodegradation occurs naturally or can be enhanced via nutrient amendments and/or aeration, and can occur under aerobic or anaerobic groundwater conditions. Aerobic BTEX degradation is common, and molecular oxygen is readily used by microorganisms to degrade BTEX compounds. However, even in aerobic groundwater, degradation may be via anaerobic biodegradation pathways due to rapid consumption of available oxygen within the more contaminated portion of the plume. Degradation of BTEX compounds by microorganisms in anaerobic aquifers has been noted by several researchers. BTEX degradation has been reported under nitrate-reducing conditions (Anid et al. 1993; Zeyer et al. 1986), sulfate-reducing conditions (Edwards et al. 1992), and methanogenic conditions (Grbic-Galic & Vogel 1987). Physical and chemical conditions within the aquifer, along with the number, type, and physiological state of the microorganisms present affect degradation. In this paper, we discuss the results of a study of the physicochemical and biological parameters associated with a BTEX plume in a sulfate-rich anaerobic aquifer.

EXPERIMENTAL PROCEDURES AND MATERIALS

A BTEX plume has developed in sulfate-rich anaerobic groundwater in Perth, Western Australia. The plume extends over 400 m downgradient from the leakage point, and is approximately 30 m wide and 0.5 to 2 m thick along most of its length. Further details of the plume are given in Davis et al. (1993).

Ten multiport monitoring bores (denoted MP) were installed along the length of the plume, and a control bore (MP9) was installed upgradient of the contaminant leakage point (Figure 1). Each monitoring bore consists of 12 mini-piezometers. Construction details for the mini-piezometers are given in Davis et al. (1992). Groundwater samples were recovered from the mini-piezometers by suction using sterile glass syringes to avoid volatile losses and contact with the atmosphere. Samples for chemical and microbial tests were collected on the same sampling occasion to enable comparison of results.

Sample analyses were carried out by atomic adsorption spectroscopy for iron, colorimetric tests for hydrogen sulfide and sulfate, and titrimetric analysis for bicarbonate. Samples taken for analysis for hydrocarbon compounds were extracted in the field using a microextraction method that used diethyl ether followed by analysis by gas chromatography-mass spectrometry (GC-MS) (Patterson et al. 1993). Samples collected for microbial analysis were immediately fixed with 1% formalin (final concentration), then frozen until tested.

To determine total microbial numbers (cells/mL) in groundwater, samples were diluted 1:10 in filter-sterilized Ringers solution, and 1 mL of the dilution was filtered through an Irgalan Black-stained, 25 mm, 0.2 μm polycarbonate filter (Poretics) and stained with a 10 μg/mL solution of 4',6-diamidino-2-phenylindole (DAPI) using the method of Hobbie et al. (1977). The number of blue fluorescent cells present per field in 30 fields was counted. From the average number of cells per field of view, the number of microbial cells/mL of the original sample was determined using the method described in Standard Methods for the Examination of Water and Wastewater (Clesceri et al. 1984)

RESULTS

Analysis of the organic and inorganic compounds in groundwater indicated that changes in the sulfate, bicarbonate, and hydrogen sulfide concentrations, and Eh values could be related to concentration changes of hydrocarbons in the

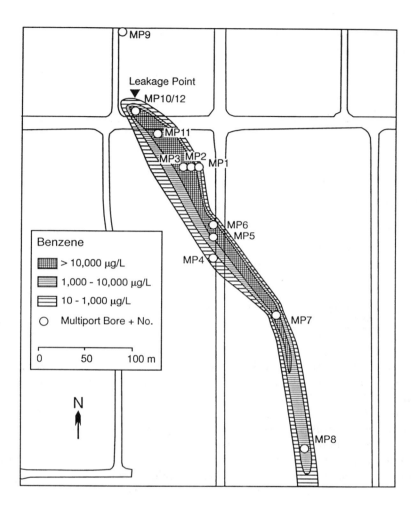

FIGURE 1. Benzene plume and multiport bore placements.

BTEX plume. Means and standard deviations for the different physicochemical parameters inside and outside the contaminant plume are given in Table 1. The data show an overall decrease in iron and sulfate concentrations, a decrease in redox potential, and an increase in carbonate and hydrogen sulfide concentrations inside the contaminant plume, compared to background data outside the contaminant plume.

Depth profiles for the groundwater parameters for selected MP bores are also illustrated in Figure 2. These results show that sulfate concentrations decreased significantly in the zone of BTEX contamination; decreasing from background concentrations of 20 to 40 mg/L to below 1 mg/L at some locations (not shown in Figure 2). The results also show that, in the presence of hydrocarbons,

TABLE 1. Chemical parameters in the contaminant plume and in background groundwater.

Parameter	Contaminant Plume		Background	
	Mean	Standard Deviation	Mean	Standard Deviation
Fe^{2+} [mg/L]	0.8	1.0	2.0	1.0
SO_4^{2-} [mg/L]	14.9	8.5	28.6	6.3
HCO_3^- [mg/L]	57.3	31.5	16.7	8.5
H_2S [mg/L]	1.2	1.3	<0.1	<0.1
EC [µS/cm]	0.72	0.16	0.87	0.14
pH	6.0	0.3	5.62	0.16
Eh [mV]	39	65	228	62
O_2%	0.5	0.1	0.6	0.1
No. of samples	24		9	

bicarbonate and hydrogen sulfide concentrations increase. This did not occur in the upgradient control bore MP9, where no significant trends could be determined, i.e., bicarbonate concentrations were low in MP9 and no hydrogen sulfide was detected. Eh values are substantially lower in the BTEX-contaminated zone, but higher in the control bore in the absence of hydrocarbons.

From these data, indicative mass balances can be calculated assuming that sulfate reduction is the principal mechanism for degradation of the hydrocarbons within the plume and taking toluene as a model hydrocarbon. The equation used to calculate the theoretical mass balances can be written as (Beller et al. 1992):

$$C_7H_8 + 4.5\ SO_4^{2-} + 3H_2O \rightarrow 2.25\ HS^- + 2.25\ H_2S + 7\ HCO_3^- + 0.25\ H^+ \quad (1)$$

From Equation 1, 1 mg of toluene reacts with 4.7 mg of sulfate and produces 1.6 mg of sulfide ions and hydrogen sulfide combined, and 4.6 mg of bicarbonate. The sulfate data for bore MP11 in Figure 2, when integrated over the vertical, indicate that sulfate concentrations have decreased by an average of 14 mg/L over the 1.25 m contaminated zone, which would imply a toluene-equivalent concentration decrease of 3 mg/L (i.e., 3,000 µg/L). This stoichiometry would also imply an average concentration increase of 4.8 mg/L for hydrogen sulfide and 14 mg/L for bicarbonate.

The total microbial cell count (cells/mL) detected in groundwater at different ports of the monitoring bores is presented in Figure 3. Cell numbers were highest near the leak of gasoline (i.e., MP12) and decreased downgradient, along the length of the plume. Bore MP12 had up to 4.6×10^6 cells/mL, while further downgradient along the plume the maximum cell count decreased to between 1.5×10^5 and 3.4×10^5 cells/mL. The microbial cell count in groundwater from each of the ports sampled in each MP bore either decreased with depth (e.g., bores MP12 and MP7) or did not change significantly (e.g., bore MP3). The cell count for samples collected from the control bore (MP9) was high (4.3×10^5) in the uppermost port, but was less than 5×10^4 cells/mL in the ports sampled at lower depths.

DISCUSSION AND CONCLUSIONS

Changes in the chemical composition of the groundwater, along with the increase in microbial cell numbers, indicate that an increase in microbial metabolic activity is occurring within the BTEX plume. The increased microbial activity is apparently manifest as reductions in the concentrations of the hydrocarbon compounds. Natural (intrinsic) rates of degradation of the BTEX compounds and naphthalene within the plume had been previously determined by Thierrin et al. (1992, 1993) through a small-scale tracer test using deuterated BTEX compounds and naphthalene as tracers. Calculated degradation rates indicated that benzene was least degraded, and that toluene, xylene and naphthalene had depth-averaged half-lives in the range of 30-230 days. These data and trends in the physicochemical parameters indicate that microbially-induced degradation of the hydrocarbons is occurring and strongly suggest that sulfate reduction is closely linked with the degradation process.

Although the tracer test demonstrated that natural degradation was occurring for toluene, *p*-xylene, and naphthalene in the groundwater and that degradation was linked to changes in sulfate, bicarbonate, and hydrogen sulfide concentrations and Eh value changes (Thierrin et al. 1993), no direct link could be established between the degradation rates and total cell numbers apart from the fact that cell numbers were elevated in the region in which the degradation study was undertaken. However, the increase in total microbial cell numbers within the plume indicate an increase in microbial activity when compared to the results of cell numbers present outside the plume. This is supported by results from a previous sampling (data not shown) which also showed elevated cell numbers within the plume. The survey of the number of microbial cells/mL along the length of the plume revealed that the cell numbers were highest close to the leakage point and showed marginal decrease downgradient (Figure 3). The number of cells/mL within the plume was always higher than the number of cells/mL detected in the lower ports of the control bore (MP9) which were less than 1×10^5 cells/mL. The unexpected high microbial numbers detected in the top port of the control bore may be attributed to influences from a school playing field adjacent to the bore (e.g., fertilizers and regular irrigation).

The number of cells/mL in groundwater sampled from each bore either decreased with depth (e.g., bores MP12 and MP7) or showed no significant change (e.g., bore MP3), however, no direct correlation could be made between

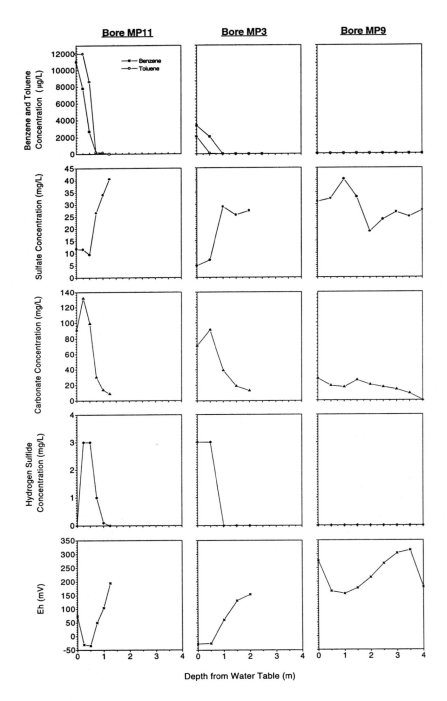

FIGURE 2. Examples of vertical changes of various chemical species in multi-port bores.

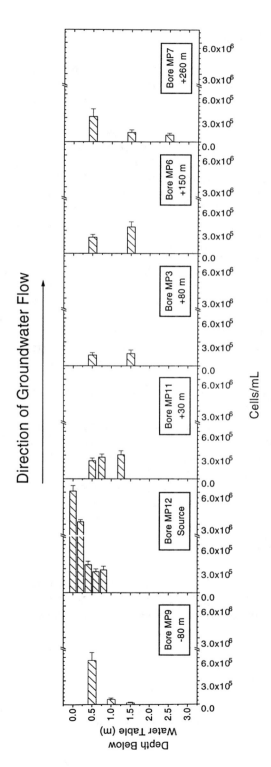

FIGURE 3. Number of cells/mL in vertical profiles of multiport bores and along BTEX plume.

the total cell numbers at specific depths at a bore and changes in the various chemical species in the contaminated groundwater. For a BTEX plume in gravel aquifers, Mikesell et al. (1991) found that the number of BTEX-degrading denitrifying bacteria varied depending on the depth from which the sample had been collected. The variation was not a gradual variation as in the present study, but occurred at a number of unrelated depths. However, Mikesell et al. (1991) found that the denitrifying bacteria were associated with BTEX degradation. The lack of direct correlation in the present study also emphasizes the need to study specific sections of the microbial community. Work is currently progressing on the study of some of the major bacterial groups (e.g., sulfate-reducing bacteria, Proteobacteria subgroups) using oligonucleotide probes and phospholipid analysis.

Mass balances modeled on the stoichiometry in Equation 1 and based on reductions in the concentration of sulfate indicated greater increases in hydrogen sulfide concentrations and lower concentrations of bicarbonate than those observed (Figure 2). The differences in sulfide concentrations may be due to the reaction of iron species found at the site with sulfide ion or with hydrogen sulfide. The higher than predicted bicarbonate concentrations (three-fold higher) could possibly indicate that other biodegradation processes, as well as sulfate reduction, are occurring in the sulfate-depleted zone. Ignoring dispersion and source variability, this is supported by total hydrocarbon concentration reductions between boreholes MP11 and MP3, which on average decrease approximately four-fold (from 22 mg/L to 5 mg/L), much greater than the 3 mg/L decrease predicted from the sulfate depletion profiles of Figure 2. Other metabolically active bacteria (perhaps iron reducers) are therefore likely to be substantial degraders of hydrocarbons in the plume.

The importance of sulfate reduction to the degradation of BTEX in the current study still remains to be fully determined. The results obtained so far, however, demonstrate that natural degradation of BTEX in the anaerobic aquifer underlying Perth is reflected in sulfate reduction and changes in several other related chemical parameters (e.g., carbonate concentrations) as well as total cell numbers, when compared with uncontaminated regions of the same aquifer.

ACKNOWLEDGMENTS

This project is funded in part by the Water Authority of Western Australia and the Australian Institute for Petroleum. The authors would like to thank Bradley M. Patterson, Michael Lambert, and Tracy Milligan for their assistance in this research.

REFERENCES

Anid, P. J., P.J.J. Alvarez, and T. M. Vogel. 1993. "Biodegradation of Monoaromatic Hydrocarbons in Aquifer Columns Amended with Hydrogen Peroxide and Nitrate." *Water Research* 27: 685-691.

Beller, H. R., D. Grbic-Galic, and M. Reinhard. 1992. "Microbial Degradation of Toluene Under Sulfate-Reducing Conditions and the Influence of Iron on the Process." *Applied and Environmental Microbiology 58*: 786-793.

Clesceri, L. S., A. E. Greenberg, and R. R. Trussell (Eds). 1989. "Direct Total Microbial Count." *Standard Methods for the Examination of Water and Wastewater*, 17th ed. pp. 9-64 - 9-66. American Public Health Association, Washington, DC.

Davis, G. B., C. Barber, D. Briegel, T. R. Power, and B. M. Patterson. 1992. "Sampling Groundwater Quality for Inorganics and Organics: Some Old and New Ideas." *International Drill Conference '92.* pp. 24.1-24.9.

Davis, G. B., C. D. Johnston, J. Thierrin, T. R. Power and B. M. Patterson. 1993. "Characterising the Distribution of Dissolved and Residual NAPL Petroleum Hydrocarbons in Unconfined Aquifers to Effect Remediation." *AGSO J. of Australian Geology & Geophysics 14*(2/3), 243-248.

Edwards, E. A., L. E. Wills, M. Reinhard, and D. Grbic-Galic. 1992. "Anaerobic Degradation of Toluene and Xylene by Aquifer Microorganisms under Sulfate-Reducing Conditions." *Applied and Environmental Microbiology 58*: 794-800.

Grbic-Galic, D., and T. M. Vogel. 1987. "Transformation of Toluene and Benzene by Mixed Methanogenic Cultures." *Applied and Environmental Microbiology 53*: 254-260.

Hobbie, J. E., R. J. Daley, and S. Jasper. 1977. "Use of Nucleopore Filters for Counting Bacteria by Fluorescence Microscopy." *Applied and Environmental Microbiology 33*: 1225-1228.

Mikesell, M. D., R. H. Olsen, and J. J. Kukor. 1991. "Stratification of Anoxic BTEX-Degrading Bacteria at Three Petroleum-Contaminated Sites." In R. E. Hinchee and R. F. Olfenbuttel (Eds.), *In Situ Bioreclamation: Applications and Investigations for Hydrocarbon and Contaminated Site Remediation*, pp. 351-362. Butterworth-Heinemann, Stoneham, MA.

Patterson, B. M., T. R. Power, and C. Barber. 1993. "Comparison of Two Integrated Methods for the Collection and Analysis of Volatile Organic Compounds in Ground Water." *Ground Water Monitoring and Remediation Summer 1993*:118-123.

Thierrin, J., G. B. Davis, C. Barber, and T. R. Power. 1992. "Use of Deuterated Organic Compounds as Groundwater Tracers for Determination of Natural Degradation Rates within a Contaminated Zone." In H. Hötzl and A. Werner (Eds.), *Tracer Hydrology*, pp. 85-91. A. A. Balkema, Rotterdam.

Thierrin, J., G. B. Davis, C. Barber, B. M. Patterson, F. Pribac, T. R. Power, and M. Lambert. 1993. "Natural Degradation Rates of BTEX Compounds and Naphthalene in a Sulphate Reducing Groundwater Environment." *Hydrological Sciences Journal 38*: 309-322.

Zeyer, J., E. P. Kuhn, and P. Schwarzenbach. 1986. "Rapid Microbial Mineralization of Toluene and 1,3-Dimethylbenzene in the Absence of Molecular Oxygen." *Applied and Environmental Microbiology 52*: 944-947.

Natural Attenuation of Xenobiotic Compounds: Anaerobic Field Injection Experiment

Kirsten Rügge, Poul L. Bjerg,
Hans Mosbæk, and Thomas H. Christensen

ABSTRACT

Currently, a continuous field injection experiment is being performed in the anaerobic part of a pollution plume downgradient of the Grindsted Landfill in Denmark. This natural gradient experiment includes an injection of 18 different xenobiotic compounds with bromide as a tracer. The injection is taking place under methanogenic/sulfate-reducing conditions and the compounds will, as they migrate with the groundwater, pass through a zone where the redox conditions have been determined as iron-reducing.

INTRODUCTION

In 1992 and 1993, detailed investigations were performed on the hydrogeology, redox conditions, and organic leachate plume at the Grindsted Landfill (Bjerg et al. 1995; Rügge et al. 1995). Approximately 60 m from the landfill border, most of the xenobiotic organic compounds are no longer detectable. Because dilution and sorption apparently cannot account for the disappearance of the xenobiotic compounds, it is proposed that the majority of the xenobiotic compounds in the leachate is degraded under methanogenic/sulfate-reducing or iron-reducing conditions in the aquifer. The redox conditions are identified by the distribution of the redox-sensitive compounds: CH_4, S^{2-}, SO_4^{2-}, Fe^{2+}, Mn^{2+}, NO_2^-, NO_3^-, and dissolved organic matter (measured as nonvolatile organic carbon, NVOC). On the basis of these investigations, an anaerobic field injection experiment has been initiated to establish direct proof of the apparent microbial/chemical degradation in the plume. The aim of the experiment is to provide data from the field on the degradation and degradation rates for the investigated xenobiotic compounds under anaerobic conditions. Such information is important in risk assessment of organic chemicals from waste disposal sites and from spills in anaerobic environments.

MATERIAL AND METHODS

The landfill is located on a glacial outwash plain. The injection is taking place in an aquifer of 5 to 7 m of glaciofluvial sand (Quaternary period), which is composed of medium- and coarse-grained sand and gravel. The geometric mean of the hydraulic conductivity in the aquifer is approximately 4.6×10^{-4} m/s as investigated by means of mini slug tests, with a variance of the lognormalized hydraulic conductivity of 0.47, indicating a fairly homogeneous aquifer (Bjerg et al. 1995). The average hydraulic gradient in the injection area is approximately 1.3% and the porosity of the aquifer is 0.30 to 0.35.

Six injection wells have been installed approximately 15 m downgradient of the landfill border. The width of the injection plume is 1.5 m and the depth is 1 m. The injection takes place at 4.5 to 5.5 m below ground surface (approximately 3 to 4 m below the groundwater table). Downgradient of the injection wells, a dense network of multilevel samplers have been installed. The fences of samplers have been placed 1 to 5 m apart, with transversal horizontal spacing of the samplers of 0.5 to 1.0 m dependent on the distance from the injection wells. The vertical spacing of the sampling points is 0.25 m. Eight months after start of the injection, fences of multilevel samplers are installed up to a distance of 30 m from the injection wells. The full extent of the site is planned to be 50 m, and more fences will be installed over the next months as the plume migrates downgradient. A total of approximately 150 multilevel samplers (630 sampling points) will be installed to monitor the plume as it migrates from a zone where methanogenic/sulfate-reducing conditions are supposed to prevail through a zone where indications of iron-reducing conditions exist (Bjerg et al. 1995).

To monitor the redox conditions, groundwater is sampled and analyzed for CH_4, S^{2-}, SO_4^{2-}, Fe^{2+}, Mn^{2+}, NO_2^-, NO_3^-, and NVOC every 6 to 8 weeks. Variations in the groundwater table are measured weekly in 17 piezometers placed just around the injection field and by a data logger upgradient of the injection wells. The general groundwater flow direction is monitored by groundwater table measurements in 80 piezometers every 6 to 8 weeks.

The continuous injection of the xenobiotic compounds was started on July 21, 1994. The injection solution consists of seven aromatic hydrocarbons (benzene, toluene, ethylbenzene, *o*-, *m*-, and *p*-xylene, and naphthalene), five nitroaromatic hydrocarbons (nitrobenzene, 2-CH_3-nitrobenzene, 4-CH_3-nitrobenzene, 1,2-dinitrobenzene, and 1,3-dinitrobenzene), four chlorinated aliphatic hydrocarbons (perchloroethylene, trichloroethylene, tetrachloromethane, and 1,1,1-trichloroethane), and two pesticides (Mecoprop and atrazine). These compounds are injected continuously along with bromide, which is used as tracer. A stock solution in distilled, anaerobic water (1.5 to 7 mg/L of the organic compounds and 2 g/L of bromide) is injected into the injection wells, where it is mixed with the leachate-affected groundwater. The amount of injected water in the natural gradient experiment is only 5% of the groundwater flux passing the injection wells. Thus, concentrations of the individual organic compounds of 75 to 350 µg/L and of bromide of approximately 100 mg/L are obtained just downgradient of the injection wells. Because benzene degradation under strongly anaerobic conditions is highly disputed, benzene is injected as ^{14}C-labeled to improve the precision of the analytical determination, and analysis is made of $^{14}CO_2$ to prove complete mineralization.

The continuous injection of the xenobiotic compounds and the tracer lasted for a total of 6 months as planned, and the field will be monitored for another 12 months after the end of the injection. By 8 months after the start of injection, the plume had reached a distance of approximately 25 m from the injection wells.

PRELIMINARY RESULTS

The interpretation of the experiment will be based partly on breakthrough curves from selected sampling points sampled weekly and partly on snapshots taken two or three times after the end of the injection. From the breakthrough curves, dilution, sorption, and degradation can be determined and actual degradation rates can be calculated. The snapshots provide an evaluation of the loss of the mass of solute in the system and the migration pattern of the plume in the aquifer, but will not be discussed further in this paper.

Redox Conditions

Redox samples have now been obtained and analyzed four times during the experiment. The results indicate that the injection is taking place under methanogenic/sulfate-reducing conditions. The injection of the compounds does not seem to affect the redox conditions, as no changes in the redox-sensitive parameters have been observed over time. An example of the water quality in the injected plume close to the injection wells is shown in Table 1. Reduced species as methane and iron(II) dominate, but neither nitrate nor oxygen is present.

TABLE 1. Concentrations of the redox-sensitive compounds in the plume close to the injection wells.

Parameter	Unit	Observed Range
Methane	mg CH_4/L	15-25
Sulfide	mg S^{2-}/L	0-0.5
Sulfate	mg SO_4^{2-}-S/L	0-3
Iron (II)	mg Fe^{2+}/L	100-150
Manganese (II)	mg Mn^{2+}/L	2-6
Nitrite	mg NO_2^-/L	<0.1
Nitrate	mg NO_3^-/L	<0.5
NVOC	mg C/L	50-70

Transport of Bromide

In Figure 1, breakthrough curves for bromide are shown in selected sampling points downgradient of the injection wells. At 1 m distance, the breakthrough (defined as 50% of maximum concentration) is seen after approximately 18 days; maximum concentration is reached after approximately 40 days, corresponding to a groundwater flow velocity of approximately 22 m/y depicted by bromide. Thereafter, a stable level of bromide is reached. From 1 to 2 m distance, the average groundwater flow velocity is approximately similar; at 5 m distance, the breakthrough is seen after approximately 100, days indicating slower flow velocities between 2 and 5 m from the injection. The results of the tracer at different sampling points show a groundwater flow velocity of 10 to 60 m/y (data not shown). Vertical, transversal, and longitudinal variations in flow velocities have been observed, which is to be expected from the nature of the deposits. However, this will, of course, complicate interpretation of the breakthrough curves.

Xenobiotic Compounds

In Figure 2, breakthrough curves for benzene, naphthalene, and bromide are shown for a selected sampling point in the central part of the plume 2 m from the injection wells. Relative to bromide, benzene and naphthalene move with a flow velocity of 80%, indicating that the compounds are only slightly sorbed onto

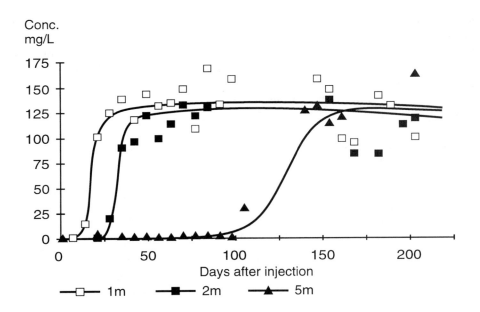

FIGURE 1. Breakthrough curves for bromide 1 m, 2 m, and 5 m from the injection wells. The sampling points are in the central part of the plume.

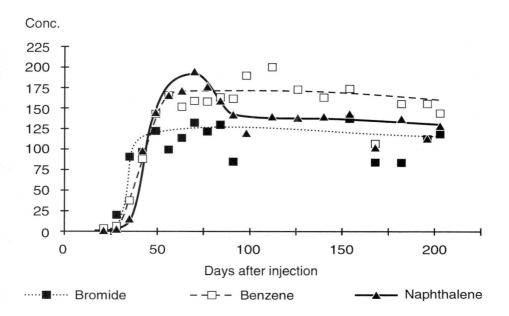

FIGURE 2. Breakthrough curves for bromide, benzene, and naphthalene 2 m from the injection wells. The sampling point is in the central part of the plume. The concentration of bromide is in mg/L, whereas the concentration of benzene and naphthalene is in µg/L.

the aquifer material. This is in accordance with our expectations, because the content of solid organic carbon is very small (f_{oc} ~0.03%). At longer distances, the relative velocities can be determined more precisely, because the dispersion will have broadened the breakthrough curves.

The breakthrough curve for benzene in Figure 2 does not indicate any degradation. However, the apparent decrease in the naphthalene concentration may be explained by degradation; further observations, both at this distance from the injection wells and further downgradient, are needed to determine whether this is the case. By 8 months after the start of injection, no significant degradation of any other aromatic compounds had been observed. Figure 3 shows the measured concentration of nitrobenzene and the estimated concentration of nitrobenzene (calculated on the basis of the injected mass of nitrobenzene and the measured concentration of bromide) 1 m from the injection wells. It is evident from this figure that nitrobenzene is degraded instantaneously as it enters the aquifer. Also degradation of tetrachloromethane, Figure 4, is observed, whereas none of the other aliphatic compounds are degraded. The degradation of nitrobenzene and tetrachloromethane under methanogenic conditions is consistent with the results presented by Nielsen et al. (1994). The plume is still being monitored, and further results are expected within the next months.

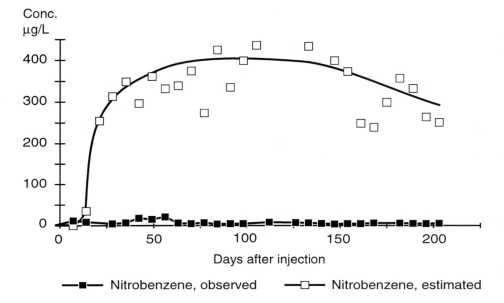

FIGURE 3. Observed and calculated concentrations of nitrobenzene 1 m from the injection wells.

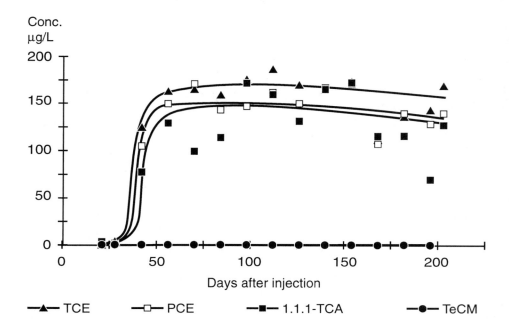

FIGURE 4. Breakthrough curves for perchloroethylene, trichloroethylene, tetrachloromethane, and 1,1,1-trichloroethane 2 m from the injection wells.

REFERENCES

Bjerg, P. L., K. Rügge, J. K. Pedersen, and T. H. Christensen. 1995. "Distribution of redox sensitive groundwater quality parameters downgradient of a landfill (Grindsted Denmark)." *Environmental Science and Technology.* In press.

Nielsen, P. H., H. Bjarnadóttir, P. L. Winter, and T. H. Christensen. 1994. "In situ and laboratory studies on the fate of specific organic compounds in an anaerobic landfill leachate plume. II: Fate of aromatic and chlorinated aliphatic compounds." Accepted for publication in *Journal of Contaminant Hydrology.*

Rügge, K., P. L. Bjerg, and T. H. Christensen. 1995. "Distribution of organic compounds from municipal solid waste in the groundwater downgradient of a landfill (Grindsted, Denmark)." *Environmental Science and Technology.* In press.

Geomicrobial and Geochemical Redox Processes in a Landfill-Polluted Aquifer

Liselotte Ludvigsen, Gorm Heron,
Hans-Jørgen Albrechtsen, and Thomas H. Christensen

ABSTRACT

The distribution of different dominant microbial-mediated redox processes in a landfill leachate-polluted aquifer (Grindsted, Denmark) was investigated. The most probable number method was utilized for detecting bacteria able to use each of the electron acceptors, and unamended incubations were utilized to detect the activity of the redox processes using the investigated electron acceptors. The redox processes investigated were methane production and reduction of sulfate, Fe(III), Mn(IV), and nitrate. The presence of methanogenic bacteria and methanogenic activity were observed close to the landfill. Sulfate-reducing bacteria and sulfate reduction were observed in the area where methanogenesis took place, but were also observed further downgradient in the leachate plume. Fe(III)-reducing bacteria were found in almost all samples from the entire anaerobic leachate plume, but no ongoing Fe(III)-reduction was observed. Sediment analysis with respect to iron species supports these findings, because no clear zone of Fe(III) depletion has been found in the leachate plume. Nitrate-reducers were found in a large section of the plume, but denitrification was observed only in the outskirts of the plume where nitrate was present.

INTRODUCTION

Leachate from municipal landfills is rich in dissolved organic matter and inorganic species that may leach into groundwater aquifers. The interaction of the highly reduced leachate with the more oxidized aquifer results in a leachate pollution plume. Depending on the composition of the leachate and redox-buffering capacity of the aquifer, a sequence of redox zones will develop (Bjerg et al. 1995), and this sequence seems to be important for controlling the degradation of organic pollutants leaching out from the landfill (Rügge et al. 1995). The identification of the redox zones at the Grindsted Landfill was based on dissolved redox-sensitive compounds and suggested the presence of a methanogenic/ sulfate-reducing zone close to the landfill, followed by iron-, manganese-, nitrate-reducing, and aerobic zones further away from the landfill (Bjerg et al. 1995). The groundwater composition indicates the presence of redox zones, but does not show whether these redox processes really are ongoing in the aquifer. Therefore, the aim of this study was to investigate the microbial processes and the geochemical factors controlling the development of the redox zones in the shallow leachate-polluted aquifer downgradient of Grindsted Landfill. Besides incubating unamended sediment samples to demonstrate the presence of the different redox processes, we focused on enumerations of the bacteria capable of mediating these different microbial redox processes in the aquifer: methanogenesis; sulfate reduction; and iron-, manganese- and nitrate-reduction. The presence or absence of the different groups of microorganisms could control the ongoing redox processes. The geochemistry of the aquifer material in terms of reactive pools of iron, manganese, and sulfur was quantified and related to the microbial processes.

MATERIALS AND METHODS

The investigated aquifer downgradient of the municipal landfill in Grindsted, Denmark, is shallow, with a groundwater level 1 to 3 m below ground surface, and consists of two geologic settings: an upper layer of medium to coarse-grained glaciofluvial sand (from the Quaternary period) and a lower layer of fine-grained micaceous sand (from the Miocene period). Corresponding sediment and groundwater samples were collected from both layers at distances from 0 to 300 m downgradient of the landfill. Typically, a group of 3 samples within a vertical distance of 1.5 m was collected from each location (Figure 1).

Groundwater samples were collected and redox-sensitive compounds were analyzed as described by Bjerg et al. (1995). The sediment samples were collected by a Waterloo piston sampler (Starr & Ingleton 1992). In the laboratory, the cores were cut into 23-cm-long sections and transferred to an anaerobic glovebox, and the outer 0.5 cm of the sediment core was pared off by a paring device slightly modified from Wilson et al. (1983). The subsamples of sediment were transferred to gastight glass jars and kept anaerobically at groundwater temperature (10°C) until analyzed. The iron- and manganese-species in the sediment samples were analyzed after wet extractions (Heron et al. 1994). Sulfur species (sulfides and pyrite attached to the sediment) were analyzed as described by Heron et al. (1994).

Bioassays were carried out in the dark at groundwater temperature (10°C) in serum bottles with a headspace of N_2-CO_2 (80 to 20%). Aquifer material was mixed with corresponding samples of groundwater in the ratio of 2 g wet weight (ww) per 3 mL of water, except for the sulfate reduction assays where the ratio was 2 g (ww) per mL. From each of the 11 sample locations, assays from one depth were performed in triplicate and one set of samples was autoclaved to serve as sterile controls. Methane production was detected by using gas chromatography to measure the accumulation of methane in the headspace of the bottle. Sulfate reduction was determined as production of $H_2^{35}S$ after addition of $^{35}SO_4^{2-}$ (final

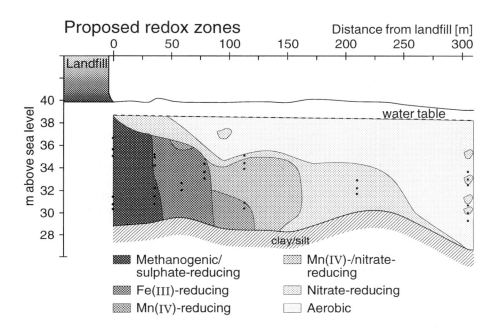

FIGURE 1. Longitudinal transect of the leachate plume in the aquifer downgradient of the Grindsted Landfill showing the 11 locations of sediment
and groundwater samples (2 or 3 samples were collected from each location). The proposed redox zones based on measured dissolved compounds
are shown (redrawn from Bjerg et al. 1995).

concentration: 8.9 MBq/L or $1.6 \cdot 10^{-8}$ g $^{35}SO_4^{2-}$/L). For each sample, four sediment suspensions were incubated for 5, 10, 15, and 20 days. The produced $H_2^{35}S$
was trapped in zinc acetate after the sample was attacked by a slightly modified
single-step chromium reduction method (Fossing & Jørgensen 1989), and quantified by scintillation counting. Fe(III)- and Mn(IV)-reduction was detected by
accumulation of Fe(II) or dissolved Mn. The Fe(II) was measured by the ferrozine
method after extraction of the suspension (0.5 M HCl for 1 h) and filtration
(0.45 µm Minisart SRP 15, Sartorius). The dissolved manganese was measured by
atomic absorption spectroscopy after filtration. Denitrification rates were measured by the acetylene blockage technique in which the accumulated N_2O in the
headspace was quantified by gas chromatography. A location was in general
interpreted as active when two out of three incubations showed significant activity of each redox process.

Most probable number (MPN) analysis was performed with aquifer solids by a
three-tube MPN method using 7-fold serial dilutions of the sediment. For each
bacteria group one MPN-test was performed from each of the 11 locations by pool-
ing three samples of aquifer solids (except for two locations where two samples
were pooled, see Figure 1). The medium used was a basal oligotrophic mineral

medium (OAM) (Albrechtsen & Christensen 1994). Electron donors were yeast extract (0.25 g/L), Bacto tryptone (Difco) (0.25 g/L), and acetate (0.2 g/L). For sulfate-reducers lactate (0.5 g/L) was also added to the medium. For methanogens and sulfate-reducers, H_2 was also added (0.8 atm). The electron acceptors were for sulfate-reducers: $CaSO_4 \cdot 2H_2O$ (0.4 g/L), $MgSO_4 \cdot 7H_2O$ (0.8 g/L) and $FeSO_4 \cdot 7H_2O$ (0.5 g/L); Fe(III)-reducers: synthetic amorphous iron oxides (ferrihydrite) (214 mg Fe(III)/L) prepared as described by Lovley & Phillips (1986); Mn(IV)-reducers: MnO_2 (Aldrich Chem. Co) (188 mg/L Mn(IV)); and nitrate-reducers: KNO_3 (35 mg/L NO_3-N). The MPN tubes were incubated at room temperature (21 to 24°C) in the dark. Tubes for methanogens, sulfate-reducers and nitrate-reducers were incubated in 2 months and tubes for Fe(III)- and Mn(IV)-reducers were incubated for 6 months to get a reliable maximum score because of the slow responds. Tubes testing for methanogens were scored as positive if the methane concentrations had raised to a level at least twice the level in the sterile controls. The presence of sulfate reducers was indicated by a black ferrous sulfide precipitate. Tubes testing for Fe(III)-and Mn(IV)-reducers were scored as positive if elevated Fe(II) and Mn(II) were observed (again, at least twice the background level). Tubes for nitrate-reducers were found positive when they were depleted in NO_3^-.

RESULTS AND DISCUSSION

A total of 31 pairs of groundwater and aquifer material samples were collected from the leachate plume downgradient of the border of the landfill (Figure 1). The groundwater compounds measured in the samples from the seven different distances downstream the landfill were in general in accordance with the distribution of the dissolved redox-sensitive parameters as described and discussed by Bjerg et al. (1995) for 160 samples from the same transect. The proposed redox zones based on the water chemistry are shown in Figure 1.

Methanogenic bacteria enumerated by MPN analysis were found in highest number in samples collected 0 and 30 m from the landfill (Figure 2). The highest number of methanogens was found in the upper layer (10^5 cells/g dw). The number of methanogens seemed to decrease with the distance, and low numbers of methanogens was found at the distances 60 m and 80 m from the landfill (10^1 and 10^2 cells/g dw). Bioassays showed that methane production took place within the first 30 m of the leachate plume. Methane production was faster in the samples from the upper quaternary layer than in the samples from the lower micaceous sand layer. The agreement between the observation of no or very low activity at the distance of 60 to 80 m from the landfill, and low numbers of methanogens in the same aquifer material, indicates that most of the methanogenic bacteria present in the aquifer are active. The observed high methane production in the plume close to the landfill is in accord with the observation of the highest level of dissolved methane in this area (Figure 3), verifying the presence of the proposed methanogenic zone.

The number of sulfate-reducing bacteria decreased with distance from the landfill. As for the methanogenic bacteria, the number of sulfate-reducing

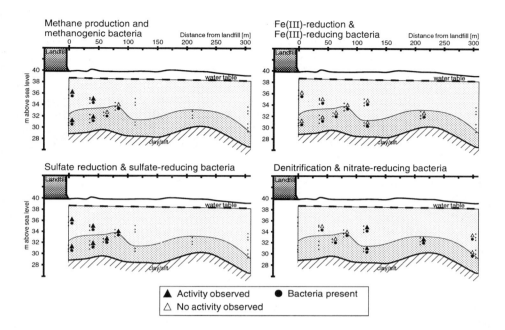

FIGURE 2. Distribution of methanogenic, sulfate-reducing, Fe(III)-reducing, and nitrate-reducing bacteria (based on 2 or 3 pooled sediment samples from each location) and activity measurement with respect to methane production, sulfate reduction, Fe(III)-reduction, and denitrification (based on an average activity measurement of 2 or 3 samples of mixed sediment and groundwater) in the plume of Grindsted Landfill. The upper Quaternary layer (white) and the lower Miocene layer (light grey) are shown.

bacteria was higher in the upper layer, decreasing with distance from 10^4 cells/ g dw to 10^3 cells/g dw in a distance of 80 m, whereas the number of sulfate-reducing bacteria in the lower micaceous layer decreased from 10^2 cells/g dw to 10^1 cells/g dw in the distance of 60 m from the landfill. Sulfate reduction was generally observed in all the same samples as where methane production was observed, but sulfate reduction was observed in a larger section of the plume within a distance of 80 m from the landfill (Figure 2). Dissolved sulfide was detected close to the landfill (cf. Heron et al. 1995), which supports our observations of ongoing sulfate reduction in this section. Potential end products from the sulfate reduction in terms of sulfides, pyrite, or organic sulfur associated with the sediment were determined in 16 of the 31 samples downstream from the landfill. In general, only low amounts of reduced sulfur were found in the sediment. However, increased contents of chromium-reducible and extractable S (S^0, pyrite, and monosulfides) were found within 30 m from the landfill (cf. Heron et al. 1995). Apparently, the geochemical data only partly support our observations of ongoing sulfate reduction.

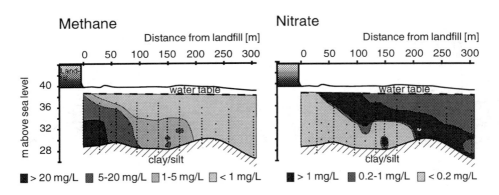

FIGURE 3. Distribution of dissolved methane and nitrate in the aquifer downgradient of the Grindsted Landfill (from Bjerg et al. 1995).

Fe(III)-reducing bacteria are present in all the investigated locations within a distance of 0 to 200 m from the landfill (Figure 2), ranging from 10^4 to 10^7 cells/g dw. Until now Fe(III)-reduction has been detected in very few of the samples and in low rates and none of the investigated locations could be defined as Fe(III)-reducing (since only one of the three samples at three locations showed Fe(III)-reduction). Even though the Fe(III)-reduction seems to be relatively rare in the plume, a great potential of Fe(III)-reduction was observed. From the incubations used in the MPN experiment (amended with amorphous Fe(III) as the only electron-acceptor and yeast extract and acetate as carbon source), a substantial potential for Fe(III)-reduction was observed as exemplified in Figure 4. At present, it is unknown why this potential of Fe(III)-reduction does not lead to detectable ongoing Fe(III)-reduction in a larger section of the aquifer. The reason could be a general natural limitation in the Fe(III) available for microorganisms in the plume or maybe due to a limitation in biodegradable organic carbon.

The observations of no substantial ongoing Fe(III)-reduction seem somewhat inconsistent with the high concentrations of dissolved Fe(II) in the section 40 to 90 m from the landfill (cf. Heron et al. 1995). The high concentrations of dissolved Fe(II) may be a result of migration with the groundwater, but ion exchange and precipitation also will affect the Fe(II) distribution in the plume. No clear Fe(III) depletion zone as a result of Fe(III)-reduction has been observed by analysis of the sediment (see Heron et al. 1995). A pool of Fe(II) bound to the sediment is found in the upper layer within 80 m from the landfill, which may be due to precipitation of Fe(II). This pool of Fe(II) may maintain the concentrations of dissolved Fe(II) long after active Fe(III)-reduction has ceased. These geochemical findings do not point out any areas of potential active Fe(III)-reduction, which indeed support our observations of the absence of microbial Fe(III)-reduction.

Mn(IV)-reducing bacteria are also distributed in a large section of the plume (data not shown). Naturally occurring Mn(IV)-reduction could be observed in the sections of elevated Mn(II) concentrations (data not shown).

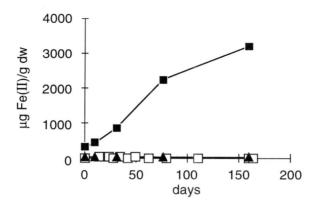

FIGURE 4. Concentrations of dissolved Fe(II) over time in an unamended suspension of sediment and groundwater (symbol □), in a pooled sediment sample in OAM amended with acetate and amorphous iron(III) oxides (symbol ■), and a sterile control of the amended incubation (symbol ▲). The sediment samples were collected from the upper quaternary layer at a distance of 120 m from the landfill.

Nitrate-reducing bacteria were found in a large section of the plume (Figure 2) ranging from 10^4 to 10^6 cells/g dw. Denitrification was observed in the two sample locations of the plume where high nitrate concentrations made the reaction possible (see Figure 3). At a distance of 300 m from the landfill, nitrate was present in the groundwater in low concentrations, but no denitrification was observed, probably because of the presence of oxygen in the groundwater. At the two sampling areas where nitrate was reduced, the potential end products from the denitrification (nitrite and dinitrogenoxide) were also found in the groundwater.

Observations of microbial methanogenesis, sulfate reduction and denitrification in general support the observed distribution of dissolved redox-sensitive compounds and the redox zones proposed by Bjerg et al. (1995). With respect to the Fe(III)-reduction, this study has shown the importance of integrating studies on water chemistry, geochemistry, and microbiology. The observations of high levels of dissolved Fe(II) in certain sections of the plume and the lack of substantial ongoing microbial Fe(III)-reduction are currently not explainable and need further study.

REFERENCES

Albrechtsen, H.-J., and T. H. Christensen. 1994. "Evidence for microbial iron reduction in a landfill leachate-polluted aquifer (Vejen, Denmark)." *Applied and Environmental Microbiology* 60: 3920-3925.

Bjerg, P. L., K. Rügge, J. K. Pedersen, and T. H. Christensen. 1995. "Distribution of redox sensitive groundwater quality parameters downgradient of a landfill (Grindsted, Denmark)." Accepted for publication in *Environmental Science and Technology*.

Fossing, H., and B. B. Jørgensen. 1989. "Measurement of bacterial sulfate reduction in sediments: Evaluation of a single-step chromium reduction method." *Biochemistry 8*: 205-222.

Heron, G., P. L. Bjerg, and T. H. Christensen. 1995. "Redox buffering in shallow aquifers contaminated by leachate." In R. E. Hinchee, J. T. Wilson, and D. C. Downey (Eds.), *Intrinsic Bioremediation*, pp. 143-151. Battelle Press, Columbus, OH.

Heron, G., C. Crouzet, A.C.M. Bourg, and T. H. Christensen. 1994. "Speciation of Fe(II) and Fe(III) in contaminated aquifer sediments using chemical extraction techniques." *Environmental Science and Technology 28*: 1698-1705.

Lovley, D. R., and E.J.P. Phillips. 1986. "Availability of ferric iron for microbially reducible ferric iron in bottom sediments of the freshwater tidal Potomac River." *Applied Environmental Microbiology 52*: 751-757.

Rügge, K., P. L. Bjerg and T. H. Christensen. 1995. "Distribution of organic compounds from municipal solid waste in the groundwater downgradient of a landfill (Grindsted, Denmark)." Accepted for publication in *Environmental Science and Technology*.

Starr, R. C., and R. A. Ingleton. 1992. "A new method for collecting core samples without a drilling rig." *Ground Water Monitoring Review 12*: 91-95.

Wilson, J. T., J. F. McNabb, D. L. Balkwill, and W. C. Ghiorse. 1983. "Enumeration and characterization of bacteria indigenous to a shallow water-table aquifer." *Ground Water 21*: 134-142.

Redox Buffering in Shallow Aquifers Contaminated by Leachate

Gorm Heron, Poul L. Bjerg, and Thomas H. Christensen

ABSTRACT

The redox conditions in two Danish landfill leachate-polluted aquifers (the Vejen and Grindsted) are discussed in terms of redox buffering. Dominant leachate contributors to reduction capacity (RDC) are dissolved organic matter and ammonium. Aquifer oxidation capacity is dominated by Fe(III) oxides, and the Vejen Landfill case shows that redox buffering by reduction of Fe(III) oxides may be important for plume development. Inorganic precipitates such as pyrite and other Fe(II) minerals may dramatically increase the oxygen demand of the aquifer. In mineral-poor aquifers such as the lower aquifer at Grindsted Landfill, redox buffering by solid electron acceptors is far less important, and smaller quantities of reduced species are formed.

INTRODUCTION

Landfill and industrial waste leachates are typically strongly reduced and contain large amounts of dissolved organic matter and reduced inorganic species (Christensen et al. 1994). The introduction of organic matter, methane, ammonium, hydrogen sulfide, and dissolved iron into an aerobic aquifer leads to redox buffering reactions. During these redox reactions, electrons donated by the reduced species in the leachate are accepted by oxidized aquifer species such as oxygen, nitrate, Mn(IV), and Fe(III) oxides and sulfate. The oxidation capacity (OXC) of an aquifer (i.e., the ability of an aquifer to resist reduction), may be expressed (modified from Barcelona and Holm 1991; Scott and Morgan 1990) as:

$$OXC = 4[O_2] + 5[NO_3^-] + [Fe(III)] + 2[Mn(IV)] + 8[SO_4^{2-}] + 4[TOC] \qquad (1)$$

assuming simple stoichiometry as given in Heron et al. (1994a). Both dissolved and solid species contribute to this OXC. The reduction capacity (RDC) of an aquifer (i.e., the ability of an aquifer to resist oxidation) may be expressed as:

$$RDC = 4[TOC] + 8[CH_4] + 8[NH_4^+] + 8[S(-II)] + 7[S(-I)] + [Fe(II)] + 2[Mn(II)] \quad (2)$$

assuming that the reduced aquifer species are oxidized to a level where free oxygen is available. Both dissolved and solid species contribute to the RDC of an aquifer volume. When the RDC of landfill leachate is calculated, only dissolved species are included.

In this study, the redox conditions in two landfill leachate plumes are discussed in the context of redox buffering. The distribution of redox-sensitive species in groundwater and aquifer solids is used to discuss the historical and ongoing processes in the plumes and the possible difficulties if enhanced in situ bioremediation is considered.

MATERIALS AND METHODS

Methods for groundwater sampling and analysis are given in Bjerg et al. (in press). Sediments were sampled anaerobically with a Waterloo piston sampler and stored in an anaerobic glovebox. Solid iron species were determined by chemical extraction methods as described by Heron et al. (1994a, 1994b) and Crouzet et al. (1995). Sediments from the Vejen plume were sampled along a central flowline (Figure 1). Sediments from Grindsted were collected at various depths along a longitudinal vertical transect downgradient of the landfill. (Sampling points appear as dots in Figure 3.)

STUDY SITES

The Vejen Landfill has leached for approximately 15 years into a shallow, glacio-fluvial aquifer dominated by oxidized quartz sand of the Weichselian period. A rising clay layer leads to almost horizontal groundwater flow downgradient of the landfill (Figure 1). The pore-flow velocity is in the order of 150 m/yr. The redox chemistry in the aquifer has previously been discussed in terms of dissolved species (Lyngkilde and Christensen 1992) and solid species (Heron and Christensen 1995).

Grindsted Landfill has leached into a shallow aquifer consisting of an upper glacio-fluvial, medium- to coarse-grained sand (Weichselian period) underlain by a fine-grained muscovite-bearing quartz sand (Figure 1). The average pore-flow velocities are 50 m/y in the upper and 10 m/y in the lower layer. The redox chemistry in terms of dissolved inorganics has been discussed by Bjerg et al. (in press).

RESULTS AND DISCUSSION

The leachate found at the border of Vejen Landfill contains more reduced species than the leachate from the border of Grindsted Landfill (Table 1). At

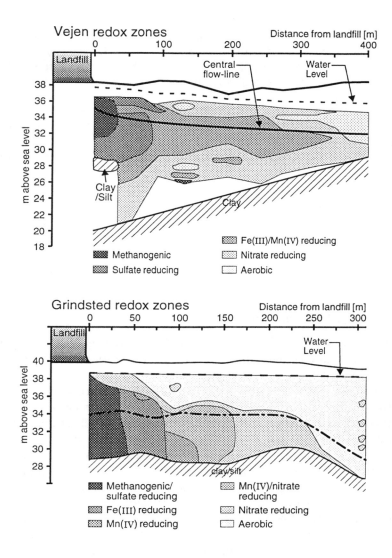

FIGURE 1. Illustration of the two studied landfill leachate plumes in terms of redox zones determined from the distribution of dissolved redox-sensitive species (modified from Lyngkilde and Christensen 1992, and Bjerg et al. in press). The boundary between the upper (Weichselian) and lower (micaceous) layer is indicated by the dotted line.

both sites, dissolved organic carbon (DOC) and ammonium are the major contributors to the RDC. The oxidation capacities of the unpolluted aquifers are quite different, with the Vejen aquifer being rich in Fe(III) oxides and the lower micaceous Grindsted layer being extremely poor in oxidized species (Table 2). In the following section, the redox buffering is discussed for each site.

TABLE 1. Major reduced dissolved species in the leachate at the border of the Vejen and Grindsted Landfills. RDC was calculated assuming simple stoichiometry and an oxidation state of 0 for the organic matter (data from Lyngkilde and Christensen 1992, and Bjerg et al. in press).

Dissolved species	Vejen Conc. (mM)	Vejen RDC (meq/L)	Grindsted, upper part Conc. (mM)	Grindsted, upper part RDC (meq/L)	Grindsted, lower part Conc. (mM)	Grindsted, lower part RDC (meq/L)
DOC	29	116	3	12	7	28
NH_4^+	13	104	4	32	7	56
CH_4	0.3	2.4	1.0	8	2	16
Fe(II)	0.1	0.1	1.0	1	2	2
Mn(II)	0.03	0.06	0.1	0.2	0.1	0.2
Total RDC		222		53		102

Vejen Landfill Leachate Plume

Fe(III) oxides are by far the largest OXC contributor in the unpolluted aquifer (Table 2), and actually this large iron pool takes part in redox processes. Figure 2 shows that within the first 200 m of the plume, the OXC related to Fe(III) and Mn(IV) oxides is depleted. During the 15 years of leaching, the original Fe(III) oxide content of 1 to 3 mg/g of sediment has been reduced. More detailed chemical extraction of the aquifer sediments reveals dramatic changes in the aquifer composition of iron species, with the precipitation of pyrite (FeS_2) and other Fe(II) species as the most significant processes (Figure 2). In the most reduced part of the plume, the pyrite precipitates (framboidal or single dipyramidal pyrites observed by scanning electron microscopy) are the major reduced species making up a large reduction capacity (Table 2). Also Fe(II) recovered in the ill-defined fraction extracted by 5M HCl contributes to the RDC. Ammonium makes up about half of the leachate RDC, and contributes about 6% of the total RDC of the reduced aquifer (Table 2).

The studies at the Vejen Landfill demonstrate the important role of Fe(III) oxides in redox buffering reactions during leachate plume development. Even crystalline Fe(III) oxides may be reduced after decades of exposure to leachate and combined biological and abiotic stress. Large amounts of reduced iron and sulfur precipitates may dramatically increase the RDC relative to the unpolluted aquifer and thereby increase the oxygen demand of the aquifer.

Grindsted Landfill Leachate Plume

The leachate strength in terms of RDC is lower at Grindsted Landfill (4-fold lower in the upper and 2-fold lower in the lower aquifer layer, Table 1) compared to Vejen. The oxidation capacity of the lower micaceous layer is very low,

TABLE 2. The estimated concentrations of oxidized species determined in the unpolluted aquifers and reduced species determined in the most reduced parts of the leachate plumes at the Vejen and Grindsted landfills. Data were evaluated from Heron and Christensen (1995) and Bjerg et al. (in press).

	Vejen		Grindsted, upper part		Grindsted, lower part	
Oxidized species in unpolluted aquifer	Conc. (mol/m³)	OXC (eq/m³)	Conc. (mol/m³)	OXC (eq/m³)	Conc. (mol/m³)	OXC (eq/m³)
O_2	0.11	0.44	0.11	0.44	0.05	0.2
NO_3^-	0.38	1.9	0.05	0.25	0.05	0.25
Mn(IV) oxides	2	4	1	2	0.2	0.4
Fe(III) oxides	57	57	30	30	2	2
SO_4^{2-}	0.15	1.2	0.08	0.6	0.08	0.6
Calculated OXC of unpolluted aquifer		65		33		3.4
Reduced species in most reduced part of plume	Conc. (mol/m³)	RDC (eq/m³)	Conc. (mol/m³)	RDC (eq/m³)	Conc. (mol/m³)	RDC (eq/m³)
Solid organic carbon	130	520	60	240	20	80
Dissolved organic carbon	10	40	1	4	2.5	10
Solid reduced S	24	336	1	14	0.05	0.7
Solid residual Fe(II)	25	25	30	30	2	2
Ion-exchangeable NH_4^+	3.2	25	1.6	12	0.8	6
Dissolved NH_4^+	4.6	36	1.4	11	2.5	20
Dissolved CH_4	0.1	0.8	0.35	3	0.7	6
Calculated RDC of most reduced zone		983		314		125

with the Fe(III) oxide content approaching the detection limit (OXC = 3 eq/m³; Table 2). The upper aquifer is similar to the Vejen aquifer except for lower contents of nitrate and sulfate, and a slightly lower Fe(III) content.

Ammonium is the dominant RDC species of the leachate (Table 1), but does not seem to be very reactive in the reduced part of the plume. The ammonium plume covers the entire anaerobic part of the plume. It is likely that ammonium is only slightly attenuated by ion-exchange processes and thus migrates through the plume until it is oxidized at the edges. This high ammonium mobility is probably due to the low cation-exchange capacity of the micaceous sand, leading to 3-fold more dissolved than ion-exchangeable ammonium in the reduced part of the aquifer (Table 2).

Apart from ammonium, dissolved organic matter is the major RDC species in the leachate, and might potentially be degraded by bacteria-using sulfate, Fe(III), Mn(IV), nitrate, or oxygen as the terminal electron acceptor, as proposed

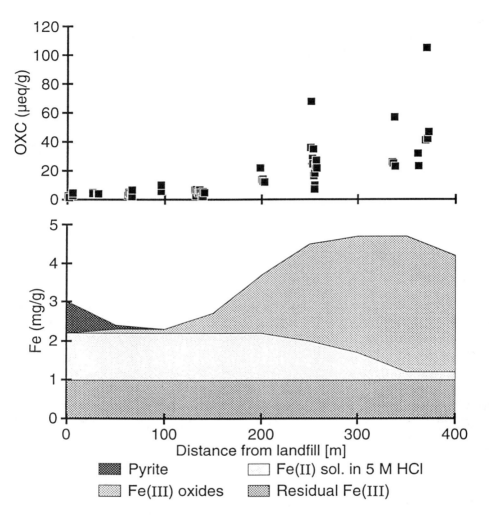

FIGURE 2. The oxidation capacity (OXC) related to iron and manganese oxides and the tentative distribution of iron species determined on aquifer solids collected along the central flowline given in Figure 1 (data from Heron and Christensen 1995).

in Figure 1. The very simple 0.5M HCl extraction shows that Fe(II) dominates among iron species extractable in weak acid in the first 50 m of the plume, especially in the lower layer. The Fe(II) found in the reduced zones either could be produced by reduction of Fe(III) oxides or simply could be the result of Fe(II) migration within the leachate plume during the leaching history of the landfill. The mixed Fe(II)-Fe(III) content of the upper reduced part of the aquifer may be due to migration of Fe(II) with the leachate or to reduction of a part of the

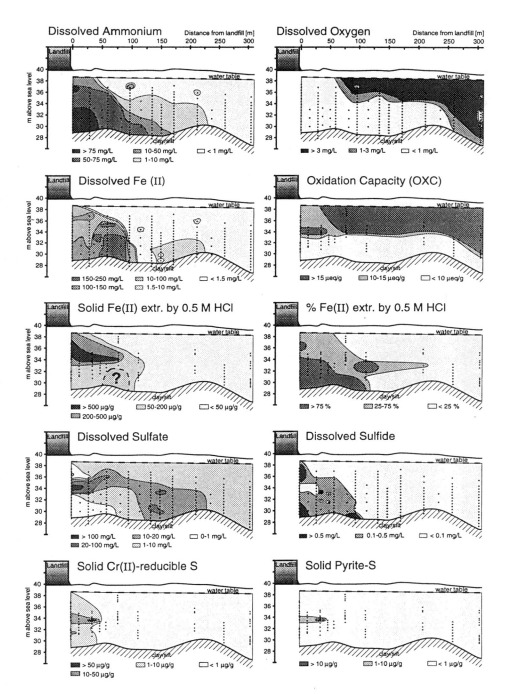

FIGURE 3. Distribution of dissolved and solid redox-sensitive species in the aquifer downgradient of the Grindsted Landfill (dissolved species from Bjerg et al. in press; solid species were determined in this study).

original Fe(III) content on the solids, because the OXC related to Fe(III) oxides is slightly decreased in the first 50 m of the plume. In any case, the redox buffering by Fe(III) oxides is far less important in the Grindsted plume than in the Vejen plume, because only small amounts of Fe(III) are shown to have reacted during the plume development.

Sulfate reduction does not seem to be a dominant redox buffering process, because only small amounts of dissolved and solid sulfides are found (Figure 3). Sulfate-rich water leaches at about 34 m above sea level and seems to migrate almost unretarded through the aquifer. Just downgradient of the land-fill border, elevated but still low solid sulfide concentrations are found. This complex distribution of sulfur species is currently being studied.

The RDC of the reduced aquifer is dominated by the solid total organic car-bon (TOC) contents (Table 2). This is in agreement with the conclusion that redox reactions involving iron and sulfur species are less important than at the Vejen site. Therefore, only low amounts of inorganic precipitates are formed.

Implications for In Situ Bioremediation

The increase in aquifer RDC and the nature of the RDC species is very rel-evant when in situ bioremediation is considered. Elevated oxygen demands of the aquifer may lead to dramatic consumption of oxidizing agents added in order to raise the redox potential or directly to oxidize the organic contami-nants. This study shows that when Fe(III) and sulfate reduction are major redox buffering processes during plume development, large amounts of reduced precipitates may be formed, and may dramatically increase the de-mand for oxidizing agents.

REFERENCES

Barcelona, M. J., and T. R. Holm. 1991. "Oxidation-reduction capacities of aquifer solids." *Environ. Sci. & Technol.*, 25, 1565-1572.

Bjerg, P. L., K. Rügge, J.K. Pedersen, and T. H. Christensen. In press. "Distribution of redox sensitive groundwater quality parameters downgradient of a landfill (Grindsted, Denmark)." *Environ. Sci. & Technol.*

Christensen, T. H., P. Kjeldsen, H.-J. Albrechtsen, G. Heron, P. H. Nielsen, P. L. Bjerg, and P. E. Holm. 1994. "Attenuation of pollutants in landfill leachate polluted aquifers." *Critical Reviews in Environ. Sci. & Technol.* 24(2): 119-202.

Crouzet, C., R. S. Altmann, and A. C. M. Bourg. 1995. "Sulfur speciation in aquifer sediments contaminated by landfill leachate: Methodology and application to the Vejen Landfill, Denmark." Submitted.

Heron, G., and T. H. Christensen. 1995. "Impact of sediment-bound iron on redox buffering in a landfill leachate polluted aquifer (Vejen, Denmark)." *Environ. Sci. & Technol.*, 29: 187-192.

Heron, G., J. C. Tjell, and T. H. Christensen. 1994a. "Oxidation capacity of aquifer sediments." *Environ. Sci. & Technol.*, 28: 153-158.

Heron, G., C. Crouzet, A. C. M. Bourg, and T. H. Christensen. 1994b. "Speciation of Fe(II) and Fe(III) in contaminated sediments using chemical extraction techniques." *Environ. Sci. & Technol., 28*: 1698-1705.

Lyngkilde, J., and T. H. Christensen. 1992. "Redox zones of a landfill leachate pollution plume (Vejen, Denmark)." *J. Contam. Hydrol., 10*: 273-289.

Scott, M. J., and J. J: Morgan. 1990. "Energetics and conservative properties of redox systems." *American Chemical Society Symposium Series, 416*: 368-378.

In Situ Bioremediation (Natural Attenuation) at a Gas Plant Waste Site

Jon S. Ginn, Ronald C. Sims, and Ishwar P. Murarka

ABSTRACT

A former manufactured gas plant (MGP) waste site in New York was evaluated with regard to natural attenuation of polycyclic aromatic hydrocarbons (PAHs). Parent-compound concentrations of PAHs within an aquifer plume were observed to decrease with time subsequent to source removal of coal tar. Biotransformation-potential studies indicated that indigenous microorganisms in soil from the site were capable of degrading naphthalene and phenanthrene. A biochemical metabolite of phenanthrene degradation, 1-hydroxy-2-naphthoic acid (1H2NA), was tentatively characterized in coal-tar-contaminated soil from the site-based on liquid chromatographic retention time. Kinetic information was developed for the disappearance of phenanthrene and 1H2NA in nonspiked contaminated soil at the site. The Microtox™ bioassay was used to evaluate toxicity trends in contaminated soil at the site. Results from the Microtox™ indicated a decreasing trend in toxicity with respect to time in contaminated site soil. Research results were evaluated with regard to the National Research Council's guidelines for evaluating in situ bioremediation, and were used to enhance site characterization and monitoring information for evaluating the role of bioremediation as part of natural attenuation of PAHs at coal-tar-contaminated sites.

INTRODUCTION

Coal-tar residues associated with MGPs are hazardous waste of particular concern because the potential carcinogenic and recalcitrant nature of various compounds, such as PAHs, which make up coal tar (Williams & Weisburger 1991). The disposal of MGP waste in pits, in lagoons, or on land often results in soil and groundwater contamination (Lee et al. 1992).

Conventional methods for remediating contaminated groundwater and soils have often relied on extractive techniques such as pumping groundwater for treatment above ground, and excavating soils for disposal, incineration, or land treatment. Pump-and-treat technologies are often limited by incomplete restoration of groundwater to health-based standards. Limitations of conventional treatment technologies have led to an increased interest in using in-place or in situ bioremediation technologies.

In situ treatment technologies (1) are generally less expensive than extractive approaches such as soil incineration (Batterman 1994); (2) treat contamination close to its source, reducing both exposure risk and cleanup costs (NRC 1993); and (3) use natural microbial and soil processes to achieve transformation, degradation, and ultimate detoxification of PAH (Dupont et al. 1988). In situ biodegradation is a natural remediation technology useful in attenuating migrating organic contaminants such as PAHs in place in the subsurface. Nonengineered in situ biodegradation may play an important role in natural attenuation: the reduction at the site of target hazardous compounds by dilution, dispersion, degradation, and/or adsorption (Tomassoni 1994).

The National Research Council (NRC 1993) has outlined a data collection and evaluation strategy to effectively demonstrate the efficacy of in situ bioremediation in reducing contaminant concentrations. This strategy consists of collecting and evaluating the following evidence: (1) documented loss of target parent compounds at a site at field scale, (2) laboratory assays demonstrating that indigenous microorganisms in site samples have the potential to transform the contaminants under expected site conditions, and (3) one or more pieces of evidence demonstrating that in situ biodegradation is actually realized in the field.

PAHs are known to degrade through a series of bacterial oxidations that subsequently transform the parent compound to various biotransformation products and, ultimately, to carbon dioxide (Boldrin et al. 1993; Guerin & Jones 1988a; Guerin & Jones 1988b; Kelley et al. 1993; MacGillivray & Shiaris 1994; Sanseverino et al. 1993). These biotransformation products have not been well characterized in complex subsurface environments containing MGP waste. The biotransformation product 1H2NA, a metabolite of phenanthrene metabolism, has been shown to accumulate in pure culture (Guerin & Jones 1988b). Successful identification, quantification, and monitoring of biotransformation products such as 1H2NA at a complex (MGP) waste site would provide evidence indicating that biotransformation was occurring. Biotransformation products could, therefore, become biomarkers or indicators of in situ bioremediation. Results reported in this paper are part of a research effort to evaluate biological processes as part of natural attenuation of PAH at a former MGP waste site containing coal-tar creosote using the NRC guidelines.

CASE STUDY FOR NATURAL ATTENUATION

During the 1960s, coal-tar residue from a former MGP site was deposited below the ground surface in a trench along a road in New York (Madsen et al.

1991a; Tayler et al. 1990). A schematic of the site, known as Site #24, is presented in Figure 1. Station #1 is located within the restored source area; station #2 is located outside the boundary of the plume and represents pristine, uncontaminated soil aquifer material; and station #19 is located near the center of the contaminated plume, approximately 200 m downgradient from station #1. The contaminated aquifer is unconfined, with the water table at about 2.5 m, and with recharge thought to occur rapidly through on-site infiltration of snow melt and rainfall (Madsen et al. 1992). The source material contained PAHs, with naphthalene reported in the highest concentrations in the source material, groundwater, and contaminated soils (EPRI 1993). Naphthalene, phenanthrene, and acenaphthalene were also found moving in the groundwater 300 to 400 m from the plume in an easterly direction (Madsen et al. 1991a). Coal-tar deposits containing dense, nonaqueous-phase liquids (DNAPLs) and light, nonaqueous-phase liquids (LNAPLs) were found only in the source area. The source material, located at station #1, was removed in late summer of 1991.

EXPERIMENTAL METHODS

The biotransformation potential of the site soil was evaluated using ^{14}C-phenanthrene radiolabeled mineralization experiments, chemical mass balance laboratory experiments, and nonradiolabeled experiments for evaluating kinetics of parent compound and 1H2NA disappearance. Preliminary laboratory experiments, indicating the disappearance of spiked nonradiolabeled phenanthrene and 1H2NA in separate reactors containing site soil, were used to design

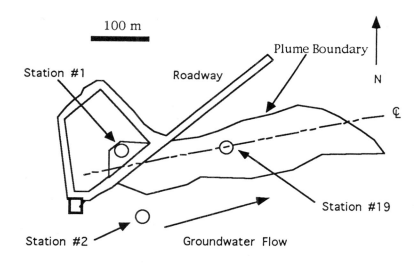

FIGURE 1. Site #24 schematic.

radiolabeled mineralization studies. Methods and results for mineralization studies are described elsewhere (Ginn et al. 1995). Chemical mass balance methods and results are also described elsewhere (Ginn et al. 1994). Results in this paper address the identification of a metabolite of phenanthrene degradation in PAH-contaminated soil, and kinetics of parent compound and 1H2NA disappearance at Site #24.

Contaminated soil from station #19 was analyzed for the presence of 1H2NA, a known microbial biochemical metabolite of phenanthrene (Guerin & Jones 1988b). Soil (50 g) was solvent extracted using a Tissuemiser and analyzed using a Shimadzu Model LC-6A high-pressure liquid chromatograph (HPLC). A mobile-phase gradient elution program, modified from the literature, was used: 90% water-10% methanol to 100% methanol at 5% min^{-1} (Guerin & Jones 1988b). The flowrate was 1.5 mL/min and the column temperature was $23 \pm 2°C$. Signal detection was accomplished with an ultraviolet (UV) detector (Shimadzu SPD-6A) with the wavelength set at 230 nm. Calibration was achieved using multiple dilutions of external standards dissolved in methanol.

Kinetic rate information was developed by determining and comparing concentrations of phenanthrene and 1H2NA in contaminated soil from station #19 (1.5 to 2.5 m depth) after incubation at 30°C and at 5°C. This temperature range covers the range of temperatures observed at the site. First-order kinetics were used to determine the rate of disappearance of each compound, to estimate a half-life value for each, and to determine a temperature correction factor for the rate of disappearance at the site.

Field measurements incorporating the Microtox bioassay were also used to evaluate toxicity trends within the contaminated plume at the site. The Microtox bioassay has been used to monitor toxicity trends during biotreatment of contaminated soils (Symons & Sims 1988; Wang et al. 1990). The Microtox bioassay is used to measure the reduction in light output of test organisms when challenged by varying concentrations of aqueous samples as an indication of the degree of toxicity of an aqueous extract of soil. The concentration (units of percent, if actual concentrations are unknown) of an aqueous extract causing a 50% reduction in light output is known as the EC50 value. EC50 values above 100% are considered nontoxic. The compound pentachlorophenol may be used as a toxic calibration standard. The EC50 value for pentachlorophenol is approximately 1 mg/L. Microtox bioassay procedures are described elsewhere (Ginn et al. 1995).

RESULTS AND DISCUSSION

Following source removal, the concentration of naphthalene in the dissolved groundwater plume was observed to dissipate in the direction of groundwater flow. Also, an inverse relationship was observed between dissolved oxygen in the groundwater and naphthalene concentration (EPRI 1993).

Oxygen was found to increase in concentration away from the centerline of the plume, while naphthalene was found to decrease in concentration away from the centerline of the plume. The inverse relationship between oxygen and naphthalene concentrations indicated that microbial degradation may be an important factor in the conversion of naphthalene, a constituent of MGP waste, at Site #24 (EPRI 1993).

Mineralization studies using spiked radiolabeled compounds in contaminated and uncontaminated soil indicated that native microorganisms at the site were capable of degrading naphthalene (Madsen et al. 1991a; Madsen et al. 1991b; Madsen et al. 1992) and phenanthrene (Ginn et al. 1995; Madsen et al. 1991a; Madsen et al. 1991b; Madsen et al. 1992), known components of coal-tar creosote. These laboratory experiments, together with field observations previously described, provided evidence that in situ bioremediation may play an important role in natural attenuation of PAHs at the site.

The biotransformation product of phenanthrene, 1H2NA, was tentatively characterized in contaminated soil from the site. A compound was identified in the solvent extract from contaminated soil taken from station #19 having the same chromatographic retention time as 1-hydroxy-2-naphthoic acid, a biochemical intermediate of phenanthrene. Referring to Figure 2, an HPLC chromatogram peak was observed to elute at approximately 21.1 min, which corresponds to the retention time of 1H2NA in a standard solution. This signal was intensified with addition of authentic 1H2NA. Background soil from station #2 was also extracted for the presence of phenanthrene and 1H2NA. Neither compound was detected in the background (control) soil.

Soil from station #19 was characterized for phenanthrene and 1H2NA. Results are presented in Table 1. Kinetic information for PAH removal from the site at station #19 was estimated by determining the concentration of phenanthrene and 1H2NA found in the site soil (1.5 to 2.5 m depth) after incubation at 30°C and at 5°C. Results are presented in Table 2. At 30°C, the half-life value of 1H2NA (7.9 days) was found to be greater than the parent compound phenanthrene by a factor of 1.8. If, during the biological degradation of phenanthrene, 1H2NA were to accumulate to quantifiable levels, then this metabolite could be used as an indicator of bioremediation at PAH-contaminated waste sites. At 5°C, 1H2NA was found to have a half-life approximately equal to the parent compound, phenanthrene.

The temperature activity coefficient theta (θ) for the degradation of phenanthrene and 1H2NA was determined using the results of the kinetic evaluation and the Arrhenius equation (Benefield & Randall 1985). Theta (θ) values for the site are shown in Table 3. These values compare favorably with values (1.0 to 1.1) reported in the literature (Tchobanoglous & Burton 1991) for biological treatment processes. Temperature activity coefficients can be used with the Arrhenius equation to correct degradation rates for seasonal temperature fluctuations, or to evaluate degradation rates at different sites with different temperatures.

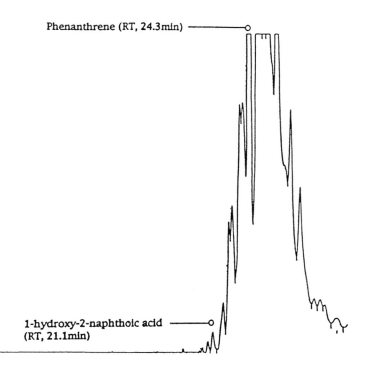

FIGURE 2. HPLC chromatogram of solvent extract of contaminated soil from
station #19. Peak retention times of 21.1 min and 24.3 min, correspond to
1H2NA and phenanthrene, respectively.

TABLE 1. Characterization of contaminated soil core from station #19.[a]

Sample Depth (m)	0.1 to 1.5	1.5 to 2.5	2.5 to 3.5
Phenanthrene	1.0	333.7	3.2
1H2NA	0.2	7.7	0.2

(a) Units for concentration values are mg/kg.

Kinetic information based on PAH disappearance in nonspiked site soil
provided an approximation of in situ PAH removal at the site. PAHs associated
with laboratory microcosms incorporating spiked compounds in soil are often
degraded rapidly, but PAHs associated with nonspiked soil matrices in field
samples show smaller decreases in concentration with time (Luthy et al. 1994).
In field samples where target compounds may reside at equilibrium in the

TABLE 2. Kinetic information for phenanthrene and 1H2NA at Site #24.[g]

PAH	Average rate, k (days⁻¹) Temp. = 30°C	Average rate, k (days⁻¹) Temp. = 5°C	Half-life estimate (days) Temp. = 30°C	Half-life estimate (days) Temp. = 5°C	Half-life values reported in literature (days)
PHEN[a]	0.15(0.01)	0.003(0.0016)	4.3	207	14[c], 2.5-26[d], 23-69[e], 200[f]
1H2NA[b]	0.08(0.03)	0.003(0.0017)	7.9	210	none reported for soil matrices

(a) PHEN = phenanthrene.
(b) 1H2NA = 1-hydroxy-2-naphthoic acid.
(c) (Bulman et al. 1987) (20°C).
(d) Range of values reviewed by (Sims & Overcash 1983).
(e) Range of values reviewed by (Dupont et al. 1988).
(f) (Coover & Sims 1987) (10°C).
(g) The standard error is shown in parentheses.

TABLE 3. Temperature-activity coefficients calculated from kinetic data.

PAH	Temperature-Activity Coefficient (θ)	Temperature-Activity Coefficient theta (θ)[b]
Phenanthrene	1.1	1.0
1H2NA	1.1	NA[a]

(a) NA not available.
(b) Calculated from Coover & Sims (1987).

subsurface, the rate of biodegradation may be influenced by coal-tar/water mass-transfer limitations for PAHs (Luthy et al. 1994), and/or the diffusion of organic compounds into and out of soil pores (Alexander & Scow 1989).

The toxicity of aqueous soil extracts from station #19 taken before source removal in 1991 was compared to the toxicity of soil extracts from station #19 taken after source removal in 1993. An EC50 value of 17.5% (toxic response) was observed for aqueous extracts of soil core material from station #19 in 1991, whereas EC50 values greater than 100% (nontoxic response) were observed for soil core material from station #19 in 1993. The Microtox bioassay indicated a trend toward decreasing toxicity with time at station #19 within the contaminated plume (Ginn et al. 1995).

CONCLUSIONS

Biological processes operating as part of natural attenuation of PAHs at a former MGP waste site in New York were evaluated with regard to the three-part strategy outlined by the National Research Council (NRC 1993). This strategy consisted of collecting and evaluating the following evidence: (1) documented loss of target parent compounds at a site at field scale, (2) laboratory assays demonstrating that microorganisms in site samples have the potential to transform the contaminants under expected site conditions, and (3) one or more pieces of evidence demonstrating that in situ biodegradation is actually realized in the field. Information was obtained for Site #24 that addressed the three criteria identified above.

With regard to the first criterion, naphthalene and phenanthrene were observed to decrease in concentration within a contaminated plume subsequent to source removal at the site (EPRI 1993).

With regard to the second criterion, laboratory microcosm studies indicated the potential of existing microorganisms at the site to be capable of degrading phenanthrene in both contaminated and uncontaminated soil from the site (Ginn et al. 1995; Madsen et al. 1991b). Presence of biotransformation products such as 1H2NA, however, were not determined in these laboratory microcosms. Future research will be directed at identifying biotransformation products generated in spiked laboratory microcosms containing contaminated and noncontaminated soil.

Regarding the third criterion, which was the focus of this paper, a microbial degradation metabolite of phenanthrene, 1H2NA, was tentatively characterized using HPLC in soil samples from the site. 1H2NA may serve as a biomarker for monitoring in situ bioremediation at Site #24. Successful identification, quantification, and monitoring of biotransformation products at complex (MGP) waste sites would provide evidence indicating that biotransformation was occurring. Biotransformation products such as 1H2NA could, therefore, become biomarkers for enhancing monitoring information concerning in situ bioremediation. The confirmation of 1H2NA in contaminated soil extracts is presently being accomplished through use of gas chromatography/ mass spectrometry.

Kinetic information developed for the disappearance of PAH in nonspiked contaminated soil provides an approximation of in situ biodegradation rates at the site. Temperature activity coefficients determined from the kinetic data developed in this evaluation compare favorably with reported values for biological reactions (Coover & Sims 1987; Tchobanoglous & Burton 1991), indicating reasonable biological activity.

The Microtox bioassay provided additional evidence in support of natural attenuation in that a trend toward decreasing toxicity within the contaminated plume was observed. The inverse relationship between oxygen and naphthalene concentrations within the plume also provided evidence in support of the role of biological processes in natural attenuation at Site #24.

ACKNOWLEDGMENTS

This work was supported by the Electric Power Research Institute (EPRI), Palo Alto, California, through a competitive contract (Agreement No. RP2879-09, "Biotransformation Studies of Organic Substances") granted to Utah State University, Logan, Utah. Dr. Ronald C. Sims was the principal investigator, and Dr. Ishwar Murarka was the EPRI Project Officer. The authors wish to thank Barbara Taylor, David Mauro, Michael Young (Meta Environmental, Inc., Watertown, Massachusetts); Eugene Madsen, Sharon Best, William Ghiorse (Division of Biological Sciences, Cornell University, Ithaca, New York); and John Ripp (Atlantic Environmental Services, Colchester, Connecticut).

REFERENCES

Alexander, M. and K. M. Scow. 1989. "Kinetics of Biodegradation in Soil." In B. L. Sawhney and K. Brown (Eds.), *Reactions and Movement of Organic Chemicals in Soils*, pp. 243-270. Soil Science Society of America, Madison, WI.

Batterman, S. 1994. "Contaminated Soil." In W. P. Cunningham, T. Ball, T. H. Cooper, E. Gorham, M. T. Hepworth, and A. A. Marcus (Eds.), *Environmental Encyclopedia*. Gale Research Inc., Detroit, MI.

Benefield, L. D., and C. W. Randall. 1985. *Biological Process Design for Wastewater Treatment*. Ibis Publishing, Charlottesville, VA.

Boldrin, B., A. Tiehm, and C. Fritzsche. 1993. "Degradation of Phenanthrene, Fluorene, Fluoranthene, and Pyrene by a *Mycobacterium* sp." *Applied and Environmental Microbiology* 59(6): 1927-1930.

Bulman, T. L., S. Lesage, P. Fowlie, and M. D. Webber. 1987. "The Fate of Polynuclear Aromatic Hydrocarbons in Soil." In J. H. Vandermeulen and S. E. Hrudey (Eds.), *Symposium of Oil Pollution in Freshwate*. Pergamon Press, Edmonton, Alberta.

Coover, M. P., and R. C. Sims. 1987. "The Effect of Temperature on Polycyclic Aromatic Hydrocarbon Persistence in an Unacclimated Agricultural Soil." *Hazardous Waste and Hazardous Materials* 4(1): 69-82.

Dupont, R. R., R. C. Sims, J. L. Sims, and D. L. Sorensen. 1988. "In Situ Biological Treatment of Hazardous Waste-contaminated Soils." In D. L. Wise (Ed.), *Biotreatment Systems*, pp. 23-94. CRC Press, Boca Raton, FL.

EPRI. 1993. "Long-Term Groundwater Monitoring Study at Site 24." *Land and Water Quality News* 7(2): 8-12.

Ginn, J. S., W. J. Doucette, and R. C. Sims. 1994. "Chemical Mass Balance Approach for Estimating Fate and Transport of Polycyclic Aromatic Metabolites in the Subsurface Environment." *Polycyclic Aromatic Compounds* 5: 225-234.

Ginn, J. S., R. C. Sims, and I. P. Murarka. 1995. "Evaluation of Biological Treatability of Soil Contaminated With Manufactured Gas Plant Waste." *Hazardous Waste and Hazardous Materials,* (in press).

Guerin, W. F., and G. E. Jones. 1988a. "Mineralization of Phenanthrene by a *Mycobacterium* sp." *Applied and Environmental Microbiology* 54(4): 937-944.

Guerin, W. F,. and G. E. Jones. 1988b. "Two-Stage Mineralization of Phenanthrene by Estuarine Enrichment Cultures." *Applied and Environmental Microbiology* 54(4): 929-936.

Kelley, I., J. P. Freeman, F. E. Evans, and C. E. Cerniglia. 1993. "Identification of Metabolites From the Degradation of Fluoranthene by *Mycobacterium* sp. Strain PYR-1." *Applied and Environmental Microbiology* 59(3): 800-806.

Lee, L. S., P. S. Rao, and I. Okuda. 1992. "Equilibrium Partitioning of Polycyclic Aromatic Hydrocarbons from Coal Tar Into Water." *Environmental Science and Technology* 26(11): 2110-2115.

Luthy, R. G., D. A. Dzombak, C. A. Peters, S. B. Roy, A. Ramaswami, D. V. Nakles, and B. R. Nott. 1994. "Remediating Tar-Contaminated Soils at Manufactured Gas Plant Sites: Technological Challenges." *Environmental Science & Technology* 28(6): 266-276.

MacGillivray, A. R., and M. P. Shiaris. 1994. "Relative Role of Eukaryotic and Prokaryotic Microorganisms in Phenanthrene Transformation in Coastal Sediments." *Applied and Environmental Microbiology* 60(4): 1154-1159.

Madsen, E. L., S. N. Levine, and W. C. Ghiorse. 1991a. "Microbiology of a Coal-Tar Disposal Site: A Preliminary Assessment." Electric Power Institute, EPRI EN-7319, Project 2879-5, Final Report.

Madsen, E. L., J. L. Sinclair, and W. C. Ghiorse. 1991b. "In Situ Biodegradation: Microbial Patterns in a Contaminated Aquifer." *Science* 252: 830-833.

Madsen, E. L., A. Winding, K. Malachowsky, T. T. Constance, and W. C. Ghiorse. 1992. "Contrasts between Subsurface Microbial Communities and Their Metabolic Adaptation to Polycyclic Aromatic Hydrocarbons at a Forested and an Urban Coal-Tar Disposal Site." *Microbial Ecology* 24: 199-213.

NRC. 1993. *In Situ Bioremediation: When Does It Work?* National Research Council, National Academy Press, Washington, DC.

Sanseverino, J., B. M. Applegate, J. M. H. King, and G. S. Sayler. 1993. "Plasmid-Mediated Mineralization of Naphthalene, Phenanthrene, and Anthracene." *Applied and Environmental Microbiology* 59(6): 1931-1937.

Sims, R. C., and M. R. Overcash. 1983. "Fate of Polynuclear Aromatic Compounds (PNAs) in Soil-Plant Systems." *Residue Reviews 88*: 1-168.

Symons, B. D., and R. C. Sims. 1988. "Assessing Detoxification of a Complex Hazardous Waste, Using the Microtox™ Bioassay." *Arch. Environ. Contam. Toxicol. 17*: 497-505.

Tayler, B., D. Mauro, M. B. Hayes, B. Holman, and M. Young. 1990. "Composition and Migration of Coal Tar-Derived Organic Compounds in a Sandy Aquifer." In *Environmental Research Conference on Groundwater Quality and Waste Disposal*, pp. 26-1 to 26-12. Electric Power Research Institute, EN-6749.

Tchobanoglous, G., and F. L. Burton. 1991. *Wastewater Engineering: Treatment, Disposal, and Reuse.* 3rd ed. McGraw-Hill Inc., New York, NY.

Tomassoni, G. 1994. *Symposium on Intrinsic Bioremediation of Ground Water,* Denver, CO. U.S. Environmental Protection Agency.

Wang, X., X. Yu, and R. Bartha. 1990. "Effect of Bioremediation on Polycyclic Aromatic Hydrocarbon Residues in Soil." *Environmental Science and Technology* 24(7): 1086-1089.

Williams, G. M., and J. H. Weisburger. 1991. "Chemical Carcinogens." In M. O. Amdur, J. Doull, and C. D. Klaassen (Eds.), *Casarett and Doull's Toxicology,* McGraw Hill, New York, NY.

Assessment of Bioremediation of a Contaminated Wetland

C. Michael Swindoll, Richard E. Perkins,
John T. Gannon, Marty Holmes,
and George A. Fisher

ABSTRACT

Evidence exists that natural remediation processes are restoring the environmental quality of the wetlands contaminated with nitrobenzene (NB) and aniline (AN). Based on our initial investigation, it appears likely that natural biodegradation is a major mechanism contributing to the wetland restoration. In situ water chemistry indicates that conditions are conducive for the growth of anaerobic microorganisms, especially sulfate- and iron-reducing bacteria. Laboratory studies show that indigenous microbes can reduce NB under sulfate-reducing, denitrifying, and methanogenic conditions. However, AN, produced as the result of NB reduction, was biodegraded only under denitrifying conditions. In situ bioremediation may be enhanced with amendments of nitrate as an electron acceptor and yeast extract as a supplemental nutrient source. Additional studies are in progress to fully elucidate the optimal conditions for in situ biodegradation of AN and NB.

INTRODUCTION

Past operational practices at a chemical manufacturing facility resulted in elevated concentrations of nitrobenzene (NB) and aniline (AN) in a wetland on the plant site. Unpublished analytical monitoring results for one area of the wetland indicate that the concentrations of these chemicals have attenuated over time. Natural biodegradation was hypothesized to be contributing to the degradation of NB and AN in the wetland. To determine the appropriate corrective response to restore the wetlands, a better understanding of the mechanisms responsible for the contaminants' attenuation was needed. This paper reports the results of the initial phase of an investigation to assess the significance of natural biodegradation to the restoration of the wetland, and to determine whether biodegradation could be enhanced through nutrient additions.

METHODS

Reconnaissance activities identified areas contaminated with NB and AN. Soil and water samples from contaminated areas were collected with a hand auger and bailer for determinations of water chemistry and for laboratory studies. The samples were transported to the laboratory in glass bottles. Laboratory microcosms, consisting of glass serum bottles containing soil slurries (10% and 20% [w/v]), were used to determine biodegradation under a variety of electron acceptor and nutrient conditions. The microcosm treatments, which included replicates and sterile controls (1 g/L mercuric chloride, or autoclaved at 121°C for 30 min at 15 to 20 psi), were prepared anaerobically and incubated in an anaerobic glove box at room temperature. Water samples for analysis were collected, reacted with anhydrous potassium carbonate (1.14 g/mL), extracted with methylene chloride, and centrifuged to break emulsions. Analytes were quantified by gas chromatography-mass spectrometry (Hewlett Packard 5890) equipped with an HP 5971A Mass Selector Detector and an HP 7673 Automated Liquid Sampler.

RESULTS AND DISCUSSION

Natural remediation processes are known to result in restoration of environmental quality at contaminated sites (Nelson 1994). In many cases, biodegradation is the most significant of the natural remediation processes. The observed decrease in surface soil concentrations of NB (from >2,000 mg/kg to <0.5 mg/kg) and AN (from >1,100 mg/kg to 180 mg/kg) over a 2-year period (unpublished data) provided evidence that naturally occurring processes were resulting in the decontamination of the wetland. Since NB and AN are known to biodegrade under aerobic (Nishino and Spain 1993; Lyons et al. 1984) and anaerobic conditions (De et al. 1994; Myers et al. 1994; Gurevich et al. 1993), it was concluded that natural biodegradation may be important to the fate of the contaminants at the site.

Water chemistry data (Table 1) provide information on in situ conditions and the impact on biodegradation. For instance, the pH of the water averaged 6.45, which is within a range favorable for microbial activity. Nitrogen, as measured as ammonia or Kjeldahl nitrogen, was at sufficient concentrations so as not to limit microbial growth. It is unknown whether sufficient phosphorus (P) is available to support biodegradation since P levels were below the 5 mg/L detection limit. The water contains relatively high amounts of sulfate and iron, which could serve as electron acceptors. The measured redox potentials were within a range where the activity of iron-reducing bacteria would be expected to predominate. The high alkalinity provided buffering capacity and conditions favorable for the growth of methanogenic bacteria. Measurements of organic compounds as dissolved organic carbon (DOC), NB, and AN show that the water contains significant amounts of organic carbon. Overall, these data

TABLE 1. Wetlands water chemical characterization data.

Parameter	Location[a]	
	1	2
Alkalinity at pH 4.5 (mg/L)	573	581
Ammonia Nitrogen (mg/L)	57	57
Kjeldahl Nitrogen (mg/L)	80	110
Nitrate (mg/L)	<0.5	<0.5
Nitrite (mg/L)	<0.5	<0.5
Orthophosphate (mg/L)	<5	<5
Sulfate (mg/L)	210	410
Sulfide (mg/L)	<8	<8
Total Iron (mg/L)	140	110
Dissolved organic carbon (mg/L)	560	495
pH	6.5	6.4
Redox (mV)	–3	–51
Dissolved oxygen (mg/L)	<0.3	—
Nitrobenzene (mg/L)	471	187
Aniline (mg/L)	189	176

(a) Samples for Locations 1 and 2 were collected at a depth of 3 to 4 ft (0.9 to 1.2 m) below ground surface in the contaminated wetlands.

indicate that in situ conditions, with the possible exception of P limitations, are conducive for natural anaerobic biodegradation.

Initial studies were designed to determine which electron acceptors could be used during NB and AN degradation. Under aerobic conditions, ^{14}C-NB and ^{14}C-AN were mineralized to $^{14}CO_2$ (unpublished data). In addition, aerobic enrichment cultures of microbes from the site could grow on NB or AN as the sole carbon sources. Although aerobic biodegradation of NB and AN was shown to occur in the laboratory study, it was concluded that aerobic processes would not be of great significance under the anoxic conditions that predominate at the site. However, aerobic biodegradation may be more important during periods when the wetlands are not flooded.

The anaerobic biodegradation of NB and AN concentrations with different electron acceptors is shown in Figures 1 and 2, respectively. Under each treatment condition, except for the sterile control, NB concentrations were reduced. Under sulfate-reducing, denitrifying, and methanogenic conditions, NB decreased within 21 days from an initial concentration of 100 mg/L to below the 0.5 mg/L analytical detection limit. In the treatment without an electron acceptor amendment, NB decreased to 9 mg/L by 103 days. NB remained unchanged at approximately 100 mg/L in the sterile control. As NB decreased, AN increased from an initial background concentration of 20 mg/L to

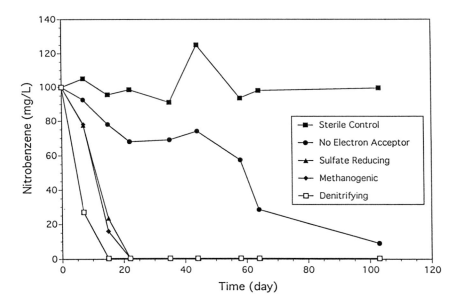

FIGURE 1.   Effect of electron acceptor on nitrobenzene biotransformation. Yeast extract (0.1% w/v) was added to the sulfate-reducing, methanogenic, and denitrifying microcosms.

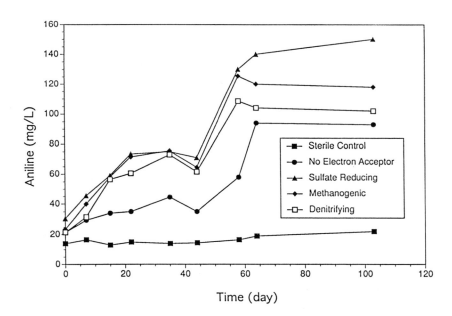

FIGURE 2.   Effect of electron acceptor on aniline production and biodegradation.   Yeast extract (0.1% w/v) was added to the sulfate-reducing, methanogenic, and denitrifying microcosms.

asymptotic levels of 90 to 140 mg/L by 64 days (Figure 2). Similar formation of AN by NB reduction has been reported (Gurevich et al. 1993). In our study, there was no indication of AN biodegradation over a 103-day period. The persistence of AN under anaerobic conditions (which could have been due to toxicity, or electron acceptor, nutrient, or other limitations), was examined in additional studies.

The changes in NB and AN concentrations under sulfate-reducing conditions and a variety of nutrient and carbon amendments are shown in Figures 3 and 4, respectively. In all the treatments, there were decreases in NB with concurrent increases in AN. However, the rates of NB biodegradation varied among different carbon amendments. In the treatment without a carbon amendment, NB degradation was slower than in treatments with carbon amendments; and NB persisted at approximately 25 mg/L after 103 days. The addition of carbon as yeast extract (YE) or sorghum (SO) resulted in the complete degradation of NB within 21 and 60 days, respectively. The addition of sucrose (SU) with YE resulted in an initial delay in NB degradation and a slower degradation than the YE and SO treatments. The addition of SU may have delayed the onset of sulfate-reducing conditions; once SU was consumed, sulfate-reducing bacteria degraded NB. The observed differences in NB biodegradation among the treatments are believed to be due to changes in microbial numbers and community structure, which occurred with the different carbon amendments. The addition of YE provided the most favorable

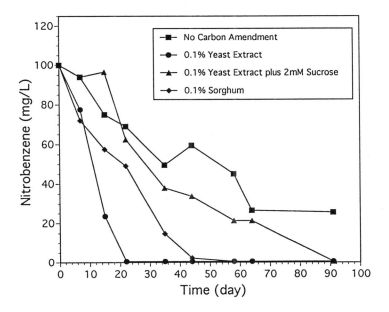

FIGURE 3. Effect of carbon source amendment on nitrobenzene biodegradation under sulfate-reducing conditions.

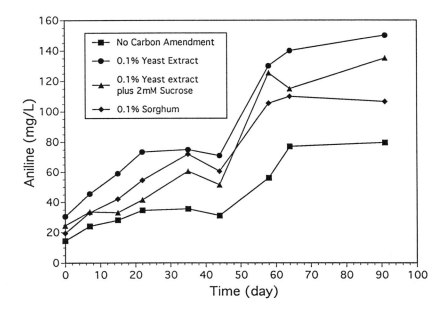

FIGURE 4. **Effect of carbon source amendment on aniline production under sulfate-reducing conditions.**

conditions for the growth of NB-degrading microbes. However, there was no evidence that AN biodegraded under any of the sulfate-reducing treatments.

Based on a report that CO_2 is required as a cosubstrate for AN biodegradation (Schnell and Schink 1991), a study of AN degradation was conducted using amendments of bicarbonate, and a range of sulfate concentrations (10, 50, and 100 mg/L as sodium sulfate). In this study, AN biodegradation was not observed under any of the treatments after 85 days. It was concluded that the cosubstrate was not the factor limiting the biodegradation of AN under sulfate-reducing conditions.

The biodegradation of NB and AN under denitrifying conditions has been examined in our recent studies. Other researchers (Myers et al. 1994; De et al. 1994) report that AN was biodegraded under denitrifying conditions following an acclimation period. Preliminary results from our studies show AN biodegradation under denitrifying conditions following an acclimation period of 16 days; AN concentrations went from averages of 44 mg/L at 16 days to <5 mg/L at 35 days. Additional studies of NB and AN biodegradation under varying denitrification conditions are in progress.

REFERENCES

De, M. A., O. A. O'Connor, and D. S. Kosson. 1994. "Metabolism of Aniline Under Different Anaerobic Electron-Accepting and Nutritional Conditions." *Environ Toxicol. Chem. 13*(2): 233-239.

Gurevich, P., A. Oren, S. C. Sarig, and Y. Henis. 1993. "Reduction of Aromatic Nitro Compounds in Anaerobic Ecosystems." *Water Sci. Technol.* 27: 89-96.

Lyons, C. D., S. Katz, and R. Bartha. 1984. Mechanisms and Pathways of Aniline Elimination from Aquatic Environments. *Appl. Environ. Microbiol.* 48(3): 491-496.

Myers, C. R., L. J. Alatalo, and J. M. Myers. 1994. "Microbial Potential for the Anaerobic Degradation of Simple Aromatic compounds in Sediments of the Milwaukee Harbor, Green Bay and Lake Erie." *Environ. Toxicol. Chem.* 13(3): 461-471.

Nelson, A.S. 1994. Passive Remediation Delivers Results. *Pollution Engineering, 94*: 40-43.

Nishino, S. and J. Spain. 1993. Degradation of Nitrobenzene by a *Pseudomonas pseudoalcaligens*. *Appl. Environ. Microbiol. 59*(8): 2520-2525.

Schnell, S. and B. Schink. 1991. Anaerobic Aniline Degradation via Reduction Deamination of 4-aminobenzoyl-CoA in *Desulfobacterium anilini. Arch. Microbiol. 155*: 183-190.

Natural Attenuation of
Coal Tar Organics in Groundwater

Mark W. G. King, James F. Barker,
and Kimberly A. Hamilton

ABSTRACT

A volume of sand containing residual coal tar creosote was emplaced below the water table under controlled field conditions to investigate natural attenuation processes for selected creosote compounds. Movement of groundwater through the source has led to the development of a complex dissolved organic plume, which has been monitored in detail for more than 1,000 days. During this period, several distinct types of behavior were evident for the monitored compounds. The *m*-xylene plume reached a maximum extent and has started to recede, while the naphthalene plume continues to migrate further from the source. Indications are that the dibenzofuran plume is at steady state, with no additional advancement and little change in plume mass over a 1-year period.

INTRODUCTION

Coal and oil processing often leads to products and wastes that are complex mixtures containing polycyclic aromatic hydrocarbons (PAHs), heterocyclic compounds, and phenolic compounds. In order to evaluate the processes affecting the natural biodegradation rate of dissolved plumes originating from these types of materials, a source of creosote was emplaced below the water table for research purposes. Creosote was used because it contains a wide range of compounds representative of many other heavy-molecular-weight, complex, hydrocarbon mixtures. Many of the compounds in creosote are reported to be aerobically biodegradable (e.g., Mueller et al. 1989). The study discussed in this technical note is part of the Ph.D. research project of the first author. It is being conducted under the Coal Tar Organics in Groundwater Program at the Waterloo Centre for Groundwater Research, and is ongoing. The purpose of this paper is to present a qualitative discussion of some preliminary data from the field component of the program and to examine some implications for complex organic plume behavior.

METHODS

The field research site is located at Canadian Forces Base (CFB) Borden, approximately 80 km northwest of Toronto, Ontario, Canada. Several hydrogeologic studies have been conducted at this location over the past 15 years. The geology and local-scale hydrogeology of the area were discussed in detail by MacFarlane et al. (1983). Mackay et al. (1986) discussed the hydrogeology and groundwater quality in the immediate area of the field experiment.

The creosote source consists of two separate lobes of creosote-contaminated sand, as shown in Figure 2, and was emplaced on August 28, 1991. The water table fluctuates within a range of approximately 0.5 m to 1.8 m above the top of the source. To install the source, sealable sheet piling was vibrated into the ground in a rectangular array (5 × 1.5 m). The sand inside the sheet piling was excavated after it was dewatered with two shallow dewatering wells. The source sand was mixed with creosote in a mortar mixer at less than 5% creosote by volume; laboratory testing had indicated that the creosote should be held effectively immobile, within the sand, at this concentration. Before addition to the sand, the raw creosote was amended with certain compounds so it would be more widely representative of a typical composition (Mueller 1989). To 69.5 kg of creosote, the following compounds were added: 0.45 kg carbazole, 0.50 kg p-cresol, and 1 kg phenol. To provide a compound representative of petroleum hydrocarbon-contaminated sites, m-xylene was also added (3 kg) . A total of approximately 74 kg of creosote was added to approximately 5,800 kg of sand.

The monitoring network of multilevel piezometers shown in Figure 1 is used for collecting groundwater samples from within and around the plume. Most of these piezometers contain 14 sampling levels with a typical vertical spacing of 20 cm and a typical depth range of from 1.8 m to 4.4 m below grade. In this technical note we discuss data from four sampling sessions that were used provide plume "snapshots" of dissolved-phase creosote compounds at four times: 55, 278, 640, and 1,008 days after source emplacement. In total, more than 4,000 samples were analyzed for the four snapshots. In addition, monthly "breakthrough" samples are collected from approximately 30 piezometers throughout the plume to verify the trends indicated by less frequent, more intensive, snapshot sampling.

The analytical methodology used for this study utilizes a solvent microextraction procedure that was developed to optimize the extraction of selected creosote compounds from groundwater, sand and pure creosote. Extracts are analyzed by gas chromatography/flame ionization detection (GC/FID).

Additional details of the source installation and analytical methodology are provided by King et al. (1994), and Malcolmson (1992). Collection and preliminary interpretation of the first two sample snapshots are reported by Nielsen and Hansen (1992). Specifications and construction details of the multilevel piezometer network used for monitoring the plume are provided by Hubbard et al. (1994).

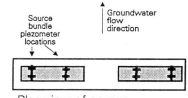

Plan view of source

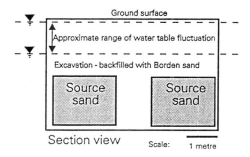

Section view   Scale: 1 metre

Source volume = 2m x 1.7m x 0.5m x 2

FIGURE 1. Plan view of groundwater monitoring network.

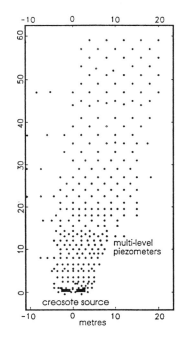

FIGURE 2. Creosote source configuration.

RESULTS

The averages of analytical results for eight samples of the altered creosote (analyzed prior to field source emplacement and after amendment with four additional compounds) are shown in Table 1. Vertically integrated plume concentrations of *m*-xylene, naphthalene, and dibenzofuran are shown in Figures 3, 4, and 5, respectively. These concentrations are calculated using the method described by Freyberg (1986), where all the point concentrations from a given multilevel piezometer are used to calculate an equivalent concentration if the total mass were to be redistributed over an arbitrary vertical aquifer thickness (in this case, 1 m). The reasons for using this method, in this case, are to assist in visualizing mass distribution and as a first step in calculating the total plume mass.

Figure 6 provides graphical plots of plume mass distribution for the three compounds for which contour plots are presented. Data for the graphical plots have been calculated by a three-step procedure. First, concentration data from each multilevel piezometer are vertically integrated (as discussed previously) to obtain a single concentration at each. Second, vertically integrated concentrations from each row of multilevel piezometers are integrated horizontally, transverse to the direction of plume migration. This procedure provides a single, integrated concentration for each row, expressed over some arbitrary distance (again, 1 m was used). Conceptually, after these two steps, the plume

TABLE 1. Initial creosote composition.

	Creosote composition (mg/kg) creosote (n = 8)
m-xylene	38,504
phenol	14,064
2,6-dimethylphenol	194
naphthalene	98,061
phenanthrene	123,703
anthracene	12,338
dibenzofuran	43,743
carbazole	3,276
pyrene	39,141
1 *m*-naphthalene	19,127
Fraction of total mass identified	0.39

All analyses performed after amendment with additional compounds (*m*-xylene, carbazole, *p*-cresol, and phenol).

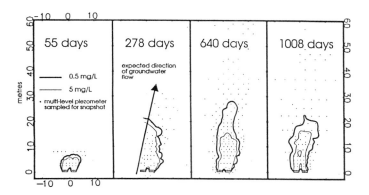

FIGURE 3. Plan view of the *m*-xylene plume from the emplaced creosote source: vertically integrated concentrations redistributed over 1-m depth.

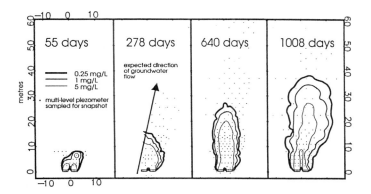

FIGURE 4. Plan view of the naphthalene plume from the emplaced creosote source: vertically integrated concentrations redistributed over 1-m depth.

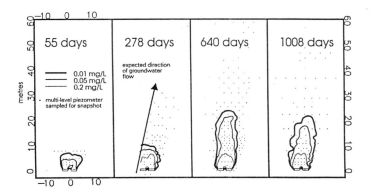

FIGURE 5. Plan view of the dibenzofuran plume from the emplaced creosote source: vertically integrated concentrations redistributed over 1-m depth.

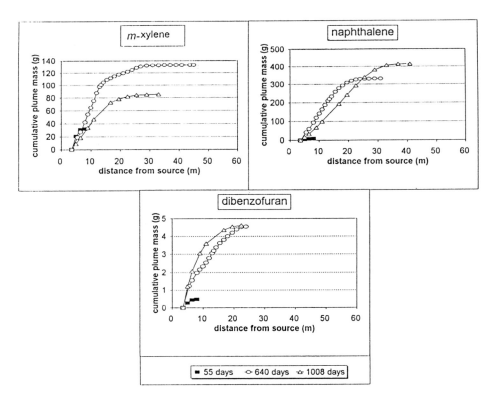

FIGURE 6. Cumulative dissolved mass for selected creosote compounds vs. distance from the emplaced creosote source; mass is calculated with successive vertical and transverse integrations of plume data; calculation of cumulative mass starts 3.5 m downgradient of source.

mass has been condensed to a volume of aquifer that is 1 m in the vertical direction, 1 m in the transverse horizontal direction, and extends downgradient from the source to the extent of the monitoring data. In the final step, cumulative mass is calculated by interpolation of concentrations from row to row along this 1-m by 1-m section, and by taking aquifer porosity into account. Besides providing information on mass distribution vs. distance from source, the plots in Figure 6 also indicate the total plume mass at a given time and the furthest detected point of migration.

DISCUSSION

Figure 3 shows that the *m*-xylene plume reached a maximum extent over the monitoring period and is now receding. It also shows that *m*-xylene is continuing to dissolve from the source. Figure 6 shows that total *m*-xylene plume mass decreased substantially from 640 days to 1,008 days, indicating that the rate of mass loss from the plume is now in excess of the rate of dissolution from the source. It also shows that the extent of detectable *m*-xylene in groundwater was closer to the source on day 1,008 than on day 640. These data suggest that migration of the *m*-xylene plume is greatly limited by some process of mass loss.

Volatilization of *m*-xylene into the vadose zone cannot account for the observed mass loss because the plume is located below the water table. Sorption onto aquifer solids is another potential sink for mass from the plume. Indeed, earlier studies at the Borden site (e.g., Hubbard et al. 1994) have shown that significant mass of *m*-xylene and other monoaromatic compounds can partition from a dissolved plume to the sorbed phase, in the Borden aquifer. However, it was also determined that this partitioning was reasonably approximated by the assumption of linear, reversible, equilibrium sorption. Consequently, a change (increase or decrease) in dissolved plume mass is expected to be associated with a change in sorbed mass that is proportional and in the same direction. In other words, an increase in sorbed mass cannot account for the observed decrease in dissolved mass. Given the above, the most likely mechanism for *m*-xylene mass loss is compound transformation, either biotic or abiotic.

The question remains as to why the *m*-xylene plume is receding instead of approaching steady state, where the rate of mass transformation equals the rate of mass flux into the plume. The relative change in magnitude of the latter process can be qualitatively evaluated by comparing the slopes of the near-source portions of the 640- and 1,008-day curves for *m*-xylene, in Figure 6. This comparison indicates that mass input has decreased over this period, probably as a result of *m*-xylene depletion from the source, which is discussed in more detail by King et al. (1994). It further indicates that the rate of *m*-xylene mass flux into the plume has decreased to below the rate of transformation within the plume and, as a consequence, the *m*-xylene plume is receding.

Plume-scale sorption data from the Borden aquifer are not available in the literature for the other two compounds (naphthalene and dibenzofuran). However, linear, reversible equilibrium sorption will also be assumed as a reasonable approximation for these compounds, given the observed behavior of *m*-xylene and the qualitative nature of the present discussion.

The naphthalene plume continues to migrate further from the source, as shown in Figure 4. Figure 6 indicates that the mass of the naphthalene plume is also increasing. If the assumption of linear, reversible equilibrium sorption is reasonable for naphthalene, then desorption of naphthalene from aquifer solids is not the cause of this increase in plume mass. Comparison of the near-source portions of the naphthalene mass distribution curves (Figure 6) indicates that, similar to *m*-xylene, the rate of naphthalene mass flux into the plume decreases from 640 to 1,008 days. However, unlike *m*-xylene, the naphthalene plume continues to advance, providing evidence that the rate of input from the source is still in excess of any mass transformation processes that may be occurring in the plume.

Data for the dibenzofuran plume, shown in Figure 5, indicate behavior that differs from the behavior of both the previous two compounds. No appreciable advancement or recession is apparent over the year that separates the last two sampling events. This observation is more easily verified by examination of Figure 6, in which the effects of variable transverse distribution are removed from the data by horizontal integration in the transverse direction. As shown, the dibenzofuran plume has approached steady-state behavior, indicated by comparable plume mass and maximum distance of plume migration between 640 and 1,008 days. Examination of the near-source portions of these curves indicates that flux of dibenzofuran into the plume is similar at the beginning and end of this period.

Review of data from monthly breakthrough samples collected from selected piezometers near the source (data not shown) indicates that the rate of mass flux is also relatively consistent throughout this period. Assuming that linear, reversible, equilibrium sorption is a reasonable approximation for dibenzofuran, then sorption of this compound to aquifer solids cannot account for the stability in dissolved mass from 640 to 1,008 days. Consequently, the observation of steady-state implies that the rate of dibenzofuran transformation in the plume is approximately equal to the rate of mass input from the source.

Additional work in progress is evaluating sorption and source dissolution to provide a more quantitative mass balance and to estimate rates of mass loss. Also under investigation are several lines of evidence to confirm that the observed mass loss is due to biodegradation. Factors affecting the rates of mass loss are also being investigated.

CONCLUSIONS

This study provides evidence that steady state of a dissolved-phase plume is achievable within a reasonable distance of an organic solute source: mass flux of dibenzofuran from the creosote source to the plume is apparently balanced by transformation of this compound in the plume. The study also highlights the potential for divergent behavior between plumes of different solutes that originate from a single complex organic source. The plume of one compound (*m*-xylene) has been observed to reach a maximum distance of plume migration and maximum mass. It is now receding and decreasing in mass, apparently due to a decrease in the rate of mass input flux to below the rate of plume transformation. The plume of a second compound (naphthalene) continues to advance and increase in mass while that of dibenzofuran has approached steady state in terms of position and mass.

ACKNOWLEDGMENTS

This work was conducted as part of the Coal Tar Organics in Groundwater Program, with funding from the Natural Sciences and Engineering Research

Council of Canada, Ontario Ministry of the Environment and a NATO Travel Grant.

REFERENCES

Freyberg, D.L. 1986. "A natural gradient experiment on solute transport in a sand aquifer 1. Spatial moments and the advection and dispersion of nonreactive tracers." *Water Resources Research* 22(13): 2031-2046.

Hubbard, C.E., J.F. Barker, S.F. O'Hannesin, M. Vandergriendt, and R.W. Gillham. 1994. *Transport and fate of dissolved methanol, methyl-tertiary-butyl-ether, and monoaromatic hydrocarbons in a shallow sand aquifer.* API Publication Number 4601.

King, M., H. Malcolmson, and J. Barker. 1994. "Groundwater plume development from a complex organic mixture." In Proceedings of *Petroleum Hydrocarbons and Organic Chemicals in Groundwater: Prevention, Detection and Remediation Conference*, pp. 411-425.

MacFarlane, D.S., J.A. Cherry, R.W. Gillham, and E.A. Sudicky. 1983. "Migration of contaminants in groundwater a landfill: a case study." *Journal of Hydrology* 63: 1-29.

Mackay D.M., D.L. Freyberg, P.V. Roberts and J.A. Cherry. 1986. "A natural gradient experiment on solute transport in a sand aquifer 1. Approach and overview of plume movement." *Water Resources Research* 22(13): 2017-2029.

Malcolmson, H. 1992. "Dissolution of an emplaced creosote source, CFB Borden, Ontario." Project Master's Report, University of Waterloo, Waterloo, Ontario.

Mueller, J.G., P.J. Chapman, and P.H. Pritchard. 1989. "Creosote contaminated sites—their potential for bioremediation." *Environmental Science and Technology* 23(10):1197-1201.

Nielsen, J.S., and S.A. Hansen. 1992. "Fate and transport of creosote compounds in a sand aquifer at CFB Borden, Ontario, Canada." Project Master's Thesis, Technical University of Denmark.

Assessment of Natural Hydrocarbon Bioremediation at Two Gas Condensate Production Sites

Gary W. Barker, Kevin T. Raterman, J. Berton Fisher,
John M. Corgan, Gary L. Trent, David R. Brown,
and Kerry L. Sublette

ABSTRACT

Condensate liquids are present in soil and groundwater at two gas production sites in the Denver-Julesburg Basin operated by Amoco. These
sites have been closely monitored since July 1993 to determine whether
intrinsic aerobic or anaerobic bioremediation of hydrocarbons occurs at
a sufficient rate and to an adequate endpoint to support a no-intervention decision. Groundwater monitoring and analysis of soil cores
strongly suggest that intrinsic bioremediation is occurring at these sites
by multiple pathways, including aerobic oxidation, Fe(III) reduction,
and sulfate reduction.

INTRODUCTION

Amoco operates more than 800 natural gas wells within the Denver Basin,
Colorado, which average about 10^5 std ft^3/d (2,830 std m^3/d) and less than
3 barrels/d associated water and condensate liquids. Structural failures of con-
crete sumps designed to contain produced water have resulted in hydrocarbon
leaks into the environment at 86 sites.

The potential cost for active remediation of these sites is conservatively
estimated at roughly $10 million. Given the remote nature of many of these
sites, real costs are likely to exceed this estimate. Amoco has sought a low-cost
alternative to active remediation wherein acceptable environmental conditions
would be restored. Natural or intrinsic bioremediation is one such option. This
option recognizes that indigenous microorganisms in the subsurface are capable of hydrocarbon degradation when critical environmental factors (e.g.,
nutrients, temperature, moisture, pH, salinity, and electron acceptors) are not
limiting. Recently, researchers have convincingly demonstrated the natural

attenuation of hydrocarbon plumes in groundwater under both aerobic and anaerobic conditions. Oxygen, nitrate, Fe(III) oxides, and sulfate have all been identified as terminal electron acceptors for the biochemical oxidation of hydrocarbons (Norris et al. 1993).

Amoco has initiated a study that seeks to determine whether intrinsic aerobic or anaerobic bioremediation of hydrocarbons occurs at the Denver Basin sites at a sufficient rate and to an adequate endpoint to support a no-intervention or intrinsic remediation option. Four tasks are specific to this objective: (1) long-term groundwater and soils monitoring (initiated in July 1993) to document field hydrocarbon losses and bioactivity over time (quarterly sampling for 2 to 3 years); (2) laboratory verification of hydrocarbon degradation by field microorganisms, and identification of primary biodegradation mechanisms (initiated in September 1993); (3) mathematical modeling to estimate biotic and abiotic losses for comparison with field observations; and (4) risk evaluation to determine potential environmental exposure pathways and anticipated losses.

Here we report the results of initial site assessments and groundwater- and soil-monitoring results to date. The implications of these data to natural attenuation of hydrocarbons at the sites are discussed.

SITE CHARACTERIZATION
AND MONITORING

In July 1993, two sites near the Platte River in agricultural areas near Fort Lupton, Colorado, were chosen for in-depth site assessments. Preliminary evaluations had shown that both soils and shallow groundwater beyond the storage-tank containment areas contained gas condensate. The potential for containment transport was deemed high. The aquifer material (gravely sands, sands, and silty sands) is highly permeable, and the water table elevations fluctuate greatly with seasonal irrigation. In addition, potential surface water receptors are near both sites. Both sites were placed in a high-priority category for further investigation.

The initial site assessment focused on hydrocarbon plume delineation. Because groundwater is shallow (3 to 5 ft [0.91 to 1.52 m]), a soil gas survey of the vadose zone could be readily conducted. Initially, a minimum of 30 vapor probes were deployed per site along the anticipated direction of groundwater flow and real-time measurements of soil gas O_2, CO_2, and volatile organic compounds (VOCs) were made.

At the KPU-2 site (Figure 1), the hydrocarbon-affected area extends from the original source within the storage impoundment area toward the north for about 150 ft (45.7 m), and the plume is narrow. Measurements of VOCs rise from background levels (100 ppm) to maximum readings (> 20,000 ppm) and return to background levels in a distance of approximately 15 ft (4.6 m). CO_2 and O_2 measurements were used as qualitative indicators of in situ bioactivity. Elevated CO_2 and diminished O_2 in soil gas concentrations coincided closely

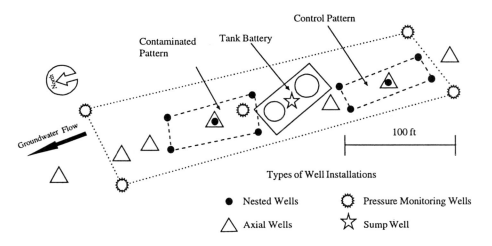

FIGURE 1. Site map of KPU-2 showing types of wells and their locations.

with the zone of abundant hydrocarbons delineated by VOC measurements (Figure 2). This is strong presumptive evidence of intrinsic hydrocarbon biodegradation at the site.

Based on the soil gas survey data, permanent groundwater monitoring wells were installed to determine the extent of hydrocarbon loss and the degree of bioactivity over time. Figure 1 shows the locations of 1-in. (2.54-cm) outer-diameter vertically nested monitoring wells that were installed in five-spot patterns within both the plume and control areas. This monitoring arrangement was adopted to define areal and vertical variations of hydrocarbon and electron-acceptor concentrations in groundwater. Each vertical-well nest consisted of three wells screened over 18-in. (45.7-cm) intervals and placed 0, 5, and 10 ft (0, 1.52, and 3.05 m) below the water table at the time of installation. At the time of installation, the water table was high. However, the water table has never been lower than just below the bottom of the topmost screened interval. Additional 2-in. (5.1-cm) outside-diameter monitoring wells were placed along the longitudinal axis of predominant groundwater flow to monitor plume migration and electron-acceptor transport. These axial wells were arranged along a path extending from upgradient of the control area, through the source, and downgradient of the plume. The 2-in. (5.1-cm) wells were screened over a 10-ft (3.05-m) interval to allow for seasonal groundwater fluctuations.

To address hydraulic modeling requirements adequately and thereby ultimately to assess the role of abiotic mechanisms (e.g., dispersion, advection) in hydrocarbon loss, both the downgradient area with hydrocarbons and the upgradient area without hydrocarbons were contained within a larger hydraulic five-spot monitoring pattern. Hydraulic monitoring wells were completed in the same way as the 2-in. (5.1-cm) monitoring wells. Pressure transducers were

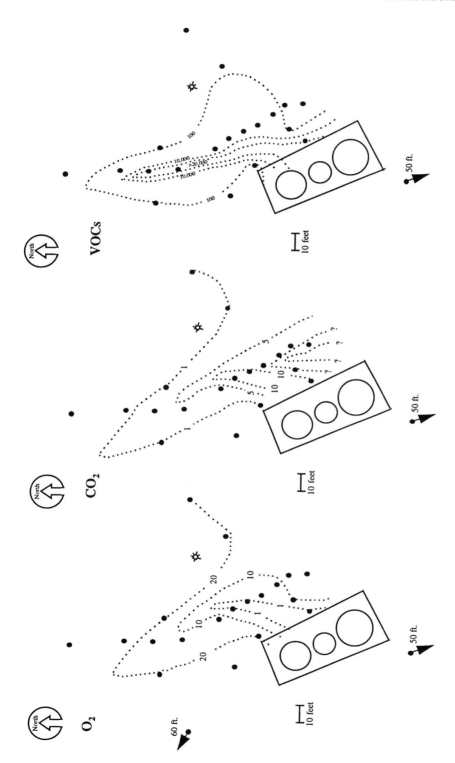

FIGURE 2. Soil gas survey data for KPU-2.

permanently installed at a fixed depth in each well. Average, maximum, and minimum water-table fluctuations are recorded daily. These measurements have a resolution of 0.5 in. (1.3 cm).

Soil cores were obtained from each site in November 1993 to document the initial soil hydrocarbon and electron-acceptor distributions. Four cores were taken from within each control and plume area. Coring locations were situated approximately halfway between the center- and corner-well clusters in each pattern quadrant. Continuous cores were obtained with a 2-in. (5.1-cm) split-spoon sampler from the surface to a total depth of 15 ft (4.6 m). Core samples were composited at 1.5-ft (45.7-cm) intervals and stored in a reduced oxygen environment at 4°C until requisite analyses could be performed.

PRELIMINARY RESULTS

Baseline groundwater samples were collected during the first week of November 1993. Water samples were obtained by producing approximately three well volumes from each monitoring well prior to sampling. Dissolved oxygen (DO), pH, and temperature were determined immediately. Individual samples were collected and analyzed within 24 h for inorganic constituents such as nitrate, sulfate, and Fe(II). Samples for benzene, toluene, ethylbenzene, and xylenes (BTEX), and total petroleum hydrocarbons (TPH) were collected in clean 40-cm^3 volatile organic analysis (VOA) vials, immediately extracted with Freon™, and shipped to Amoco's laboratory in Tulsa, Oklahoma, for analysis.

Table 1 summarizes baseline groundwater data for KPU-2. (Data from the second site show similar trends and are not reported here.) The data are organized by well depth for both control and plume volumes. Both BTEX and TPH were confined primarily to the shallow well depth within the plume volume. At this depth, median electron-acceptor concentrations were uniformly lower in the hydrocarbon plume versus the control zone, and Fe(II), a product of the utilization of Fe(III) as an electron acceptor, was higher. At the intermediate well depth, hydrocarbon concentrations were easily detectable in the plume area, but no BTEX or TPH were noted in the control samples. Trends in electron-acceptor utilization at the intermediate depth within the plume were identical to those at the shallow depths although less pronounced, presumably because of lower hydrocarbon concentrations. Deep well data indicated no appreciable hydrocarbon presence in either plume or control areas. Likewise, electron-acceptor data showed no appreciable differences. Finally, axial monitoring well data (not shown) indicated that highly soluble BTEX components had migrated a distance of only 165 ft (50.3 m) from the source over an estimated 20-year period. Sulfate and Fe(II) concentrations were at background levels at this distance. The groundwater velocity was estimated at 1.3 m/y based on a gradient of 0.4 cm/m.

Subsequent quarterly groundwater sampling events have produced similar results in terms of electron acceptor, Fe(II), BTEX, and TPH concentrations.

TABLE 1. KPU-2 baseline groundwater analyses.

	Average		Median		Minimum		Maximum	
	Control	Plume	Control	Plume	Control	Plume	Control	Plume
Shallow								
pH	6.92	6.95	7.27	6.81	6.00	6.74	7.38	7.43
Temperature,°C	10.1	9.4	10.5	9.5	8.9	8.8	10.9	9.8
Sulfate, mg/L	246.3	0	233.3	0	200.0	0	294.4	0
Nitrate, mg/L	7.7	0.2	3.1	<0.4	1.3	<0.4	19	0.6
Fe(II), mg/L	1.8	3.3	0.4	2.5	0.1	1.5	6.5	6.5
DO, mg/L	1.25	0.70	1.40	0.50	0.50	0.35	1.70	1.25
BTEX, mg/L	0.003	14.2	ND	11.6	ND	8.9	0.017	24.9
TPH, mg/L	ND	23.0	ND	20.5	ND	16.0	ND	35.0
Intermediate								
pH	5.88	6.46	5.61	6.40	5.36	6.16	7.23	6.72
Temperature,°C	10.4	11.2	10.9	11.0	9.2	10.5	11.2	12.0
Sulfate, mg/L	213.0	116.0	211.1	127.8	211.1	0.0	216.7	175.0
Nitrate, mg/L	20.2	1.9	21.7	0.8	16.8	<0.4	22.6	4.9
Fe(II), mg/L	0.1	3.2	0.05	3.0	0.05	0.3	0.1	6.0
DO, mg/L	0.49	0.45	0.40	0.38	0.40	0.30	0.75	0.60
BTEX, mg/L	ND	1.673	ND	0.51	ND	0.11	ND	4.7
TPH, mg/L	ND	2.8	ND	2.0	ND	ND	ND	8.0
Deep								
pH	6.25	6.71	5.83	6.62	5.15	6.54	7.47	7.00
Temperature,°C	10.6	11.7	10.6	11.8	9.9	10.9	11.5	12.2
Sulfate, mg/L	206.0	215.0	206.3	218.8	188.9	187.5	216.7	240.5
Nitrate, mg/L	23.2	19.0	22.6	17.7	18.2	14.2	27.5	24.4
Fe(II), mg/L	0.05	0.05	0.05	0.05	0.05	0.05	0.05	0.05
DO, mg/L	0.87	0.53	0.75	0.50	0.45	0.30	1.35	0.95
BTEX, mg/L	ND	0.010	ND	ND	ND	ND	ND	ND
TPH, mg/L	ND	ND	ND	ND	ND	ND	ND	ND

(a) At the time of sampling, the water table was at the top of the shallow screen. Each set of control or plume data is based on analyses of all wells in the respective five-spot patterns (Figure 1).

No clear decrease in BTEX or TPH concentrations has been observed to date, presumably due to replenishment of dissolved hydrocarbons from a sink of adsorbed hydrocarbons, and to water-table fluctuations.

The following observations are made on the basis of groundwater data acquired to date. The aerobic biodegradation potential of hydrocarbons appears limited because of uniformly low DO concentrations (1.4 mg/L or less) throughout the investigated area. Nitrate is present from agricultural applications of fertilizer but, because of its low concentration, has limited potential for hydrocarbon degradation. In contrast, the utilization of sulfate appears significant. Background sulfate concentrations are on the order of 230 mg/L, but within the shallow area containing hydrocarbons the sulfate concentration is 0 mg/L. Given the large initial concentrations of sulfate and its favorable stoichiometric utilization for hydrocarbon degradation, it appears that sulfate reduction may be a major means of hydrocarbon remediation at these sites. Although evidence of iron reduction exists within the contaminated area, the utilization of Fe(III) cannot be quantified from groundwater samples because of its low solubility. The solubility of Fe(II) is also low in the presence of sulfide produced by sulfate reduction.

Soil-core analyses for total iron, Fe(II), BTEX, and TPH were performed on composited samples from each 1.5-ft (45.7-cm) interval. The average and median background BTEX concentrations within the KPU-2 control volume were 0.036 and 0.02 mg/kg, respectively. Surface-soil samples typically showed higher concentrations of about 0.1 mg/kg, possibly indicating exposure to airborne BTEX. TPH values from control core samples were uniformly nondetectable (< 1 mg/kg). Comparable values within the area containing hydrocarbons for average, median, and maximum BTEX concentrations were 60.8, 0.27, and 817 mg/kg. The large difference between the average and median value reflects the vertical distribution of BTEX, which is confined to an approximate 3-ft (0.91-m) interval at the water table/air interface. Most samples outside this interval had BTEX concentrations well below 1 mg/kg. TPH values were similarly distributed with average, median, and maximum values of 379.6, nondetectable, and 4,590 mg/kg. Soil samples acquired from upgradient cores within the plume showed less BTEX and TPH than their downgradient counterparts. Because upgradient cores were situated nearest the original hydrocarbon source, this evidence supports the contention that the original source of the existing hydrocarbons in soil and groundwater has been effectively eliminated at these sites.

When hydrocarbons were present, the ratio of Fe(II) to Fe(III) in soils was increased. Assuming both iron species were initially distributed uniformly across the site, it appears that Fe(III) was subsequently reduced to Fe(II) within the zone of significant hydrocarbon presence. The reduction of iron in the presence of hydrocarbons is indicative of anaerobic biodegradation and further supports the hypothesis that intrinsic bioremediation of hydrocarbons is occurring at these sites by multiple pathways.

Finally, visual inspection of soil cores showed a significant accumulation of a black precipitate (acid volatile sulfide) associated solely with hydrocarbon presence. The accumulation of iron sulfide (FeS) in the presence of hydrocarbon is consistent with the anaerobic biodegradation of hydrocarbons by sulfate reduction.

CONCLUSIONS

At the sites investigated, intrinsic aerobic and anaerobic bioremediation play a strong role in attenuating hydrocarbons. Even though the sites are old (> 20 years), hydrocarbons are laterally and vertically confined to a small portion of the total aquifer. Also, there is strong geochemical evidence of both aerobic and anaerobic bioactivity. O_2 is much depleted and CO_2 is much elevated in the soil gas immediately overlying the groundwater associated with the hydrocarbon plumes. Soil gas immediately overlying the groundwater without hydrocarbons shows high O_2 and low CO_2 levels. Groundwater that is immediately associated with the hydrocarbon plume shows depleted levels of both nitrate and sulfate compared to groundwater not immediately associated with the hydrocarbons. Soils associated with hydrocarbons show abundant iron sulfide production. Soils not associated with hydrocarbons show no iron sulfide production. Although sulfate is depleted in groundwater associated with hydrocarbon, sulfate remains available as an electron acceptor by means of diffusion from the surrounding aquifer and by dissolution of sulfate from the solid phase. Potentially, sulfate reduction could be accelerated in the hydrocarbon-impacted areas through the addition of sulfate.

REFERENCE

Norris, R. D., R. W. Hinchee, R. Brown, P. L. McCarty, L. Semprini, J. T. Wilson, D. H. Kampbell, M. Reinhard, E. J. Bouwer, R. C. Borden, T. M. Vogel, J. M. Thomas, and C. H. Ward. 1993. *In-Situ Bioremediation of Groundwater and Geological Material: A Review of Technologies*, 68-C8-0058, R. S. Kerr Environmental Research Laboratory, Ada, OK.

Enhanced Biodegradation of Naphthalene in MGP Aquifer Microcosms

Neal D. Durant, Constance A. A. Jonkers,
Liza P. Wilson, and Edward J. Bouwer

ABSTRACT

Subsurface sediments collected from a former manufactured-gas-plant (MGP) site contain bacteria capable of mineralizing significant amounts of ^{14}C-naphthalene in aerobic (8.5 mg/L O_2) sediment-water microcosms incubated at 10°C. The extent to which electron-acceptor (O_2 and NO_3^-) and nutrient (NO_3^- and PO_4^{3-}) amendments enhanced naphthalene mineralization in these sediments varied considerably. Oxygen-amended conditions (21 mg/L O_2) resulted in the greatest rate and extent of biodegradation for most sediments. Data suggested, however, that some MGP-site sediments prefer mixed NO_3^-/O_2 electron-acceptor conditions for naphthalene biodegradation. Significant denitrification was observed in the nitrate-amended sediments exhibiting naphthalene mineralization. In most cases, PO_4^{3-} complexed with the sediments either had no effect or inhibited naphthalene mineralization. Sediments unable to mineralize naphthalene over the 6-week incubation period were characterized by low pH (< 4.0) and high SO_4^{2-} (> 500 mg/L) conditions.

INTRODUCTION

In situ bioremediation designs commonly entail the perfusion of nutrient and electron-acceptor mixtures into subsurface contaminated zones in an attempt to stimulate microbial processes. This procedure is capable of enhancing biodegradation of aqueous-phase organic contaminants, but few studies have distinguished between the relative contributions of nutrients and electron acceptors toward biodegradation enhancement. Laboratory studies investigating enhanced biodegradation of aromatic hydrocarbons in saturated aquifer sediments have observed considerable variability in the extent to which nutrients (NO_3^-, NH_4^+, and PO_4^{3-}) stimulate and/or inhibit aquifer bacteria (Swindoll et al. 1988). Determining nutritional requirements of sediment-adhered bacteria can be difficult because iron and calcium compounds present in sediments tend to complex PO_4^{3-}, thus reducing P bioavailability (Robertson and Alexander 1992). In addition, the oligotrophic behavior of many subsurface populations suggests that nutritional supplements are not a prerequisite to achieving biodegradation in aquifer sediments. Some subsurface bacteria are remarkably versatile in their ability to function both under nutrient-poor and nutrient-rich conditions (Bone and Balkwill 1988).

An ongoing research project at the Baltimore Gas & Electric (BGE) Spring Gardens Facility (a former MGP site) has been investigating the feasibility of using in situ bioremediation to contain a groundwater plume contaminated with polycyclic aromatic hydrocarbons (PAHs). Previously reported data established that active PAH-degrading populations are present in the subsurface at the site, and intrinsic bioremediation is probably attenuating plume migration (Durant et al. 1995). Current research efforts are focused on enhancing in situ biodegradation, primarily through the addition of nutrients and electron-acceptors. An objective of the present research is to determine the relative effects of NO_3^-, PO_4^{3-}, and O_2 supplements on PAH biodegradation in BGE aquifer sediments.

EXPERIMENTAL METHODS

Aquifer sediments were obtained from three boreholes using techniques described previously (Durant et al. 1995). All the samples were collected from locations exhibiting low-level groundwater contamination (naphthalene > 100 $\mu g/L$). Sediments were stored at 4°C for several months prior to use in these experiments, thus allowing time for volatilization of any sorbed organics. Ten sediments (fine- to medium-grained sands) were analyzed to measure the singular and combined effects of NO_3^- and PO_4^{3-} addition, O_2 addition, and pore-water chemistry on the aerobic mineralization of ^{14}C-naphthalene. Sediment-water microcosms were prepared in 25-mL serum vials, and consisted of 7 g sediment, 4 mL sterilized groundwater, 6.7×10^{-2} μCi ^{14}C-naphthalene, and 4 mg/L unlabeled naphthalene. The groundwater was obtained from a relatively pristine location at the BGE site, and contained < 1 mg/L NO_3^-, 10 mg/L SO_4^{2-}, and no organic contaminants. The ^{14}C-naphthalene was added using methanol as a carrier solvent, resulting in the addition of 250 mg/L methanol. To ensure oxygen availability, microcosms were prepared under aerobic conditions, and the groundwater was equilibrated with the atmosphere (attaining 8.5 mg/L dissolved O_2 before use in the microcosms). A sterile 2-mL glass vial containing 0.5 mL 1 N KOH was placed in each microcosm to trap $^{14}CO_2$ generated from mineralization of the ^{14}C-naphthalene. Microcosms were capped with Teflon™-lined butyl-rubber stoppers and aluminum crimp seals. The $^{14}CO_2$ evolved in each microcosm was determined periodically by sampling and replacing the KOH with a syringe inserted through the stopper. Radioactivity was counted by liquid scintillation with a Beckman LS 3801 liquid scintillation counter.

Biodegradation was determined as the percent ^{14}C-naphthalene mineralized to $^{14}CO_2$, less the percent of radioactivity detected in the KOH in the control microcosms. It should be noted that monitoring $^{14}CO_2$ evolution alone is a conservative means of quantifying ^{14}C-naphthalene biodegradation because a significant portion of the ^{14}C label is often incorporated into cells or transformation intermediates (Herbes and Schwall 1978). Consequently, incomplete biodegradation of naphthalene is commonly reported in sediment-water microcosm studies (Park et al. 1990; Madsen et al. 1991).

For each of the ten sediments, five sets of microcosms were prepared in triplicate, with two killed controls (1,000 mg/L NaN_3) for each set. Sediments were subjected to each of the following initial conditions: a) no nutrient amendments; b) 85 mg/L NO_3^-; c) 90 mg/L PO_4^{3-}; d) 25 mg/L NO_3^- and 90 mg/L PO_4^{3-}; and e) 21 mg/L O_2 (initial concentration) and no nutrient amendments. NO_3^- was added as $NaNO_3$ and PO_4^{3-} was added as a mixture of K-PO_4 salts. All microcosms were incubated at 10°C to simulate aquifer temperatures. Anion chemistry and pH of each microcosm's supernatant were measured at the end of the 6-week incubation to investigate how these parameters might have affected naphthalene biodegradation by the aquifer bacteria. Anions were measured on a Dionex 2010i ion chromatograph, and pH was measured with a Fisher Scientific Accumet 15 pH meter.

RESULTS AND DISCUSSION

Of the ten sediments analyzed, six exhibited significant biologic mineralization of ^{14}C-naphthalene (up to 52%) over 6 weeks of incubation (Figure 1). For most of these sediments (Borehole A, 17.4 m and 25.3 m; and Borehole B, 11.9 m and 13.7 m), microcosms amended with high initial O_2 levels (21 mg/L) yielded the greatest extent of mineralization. Since methanol (250 mg/L) was used as a carrier solvent in these experiments, it was not possible to quantify the amount of O_2 and NO_3^- consumed during naphthalene degradation. But stoichiometric calculations indicate that while there was sufficient dissolved O_2 to mineralize 4 mg/L naphthalene, the methanol exerted a large additional oxygen demand that could only be satisfied by partitioning from the microcosm headspace (Table 1). Given that the aqueous O_2 supplements significantly enhanced naphthalene mineralization, and that the initial aqueous O_2 supply was largely inadequate to satisfy methanol and naphthalene aerobic degradation requirements, it is likely that the rate and extent of naphthalene mineralization in these sediments was limited primarily by O_2 availability.

In two sediments (Borehole A, 11.6 m and 13.1 m), however, the effect of O_2 addition was less pronounced and the initial (0- to 20-day) rate of mineralization was fastest in microcosms supplemented with NO_3^- (85 mg/L) (Figure 1). Nearly complete NO_3^- consumption and evidence of nitrite production indicates that denitrification was an active process in these sediments (Figure 2), as well as in sediments receiving 25 mg/L NO_3^- and 90 mg/L PO_4^{3-} (Figure 3).

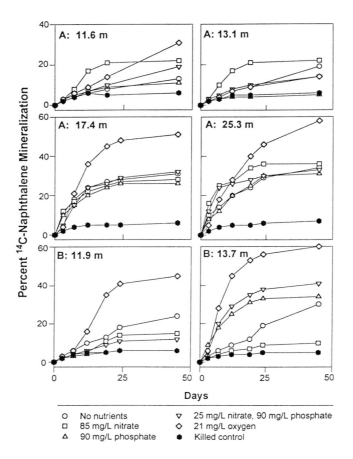

FIGURE 1. Mineralization of ^{14}C-naphthalene in sediment-water microcosms incubated in the dark at 10°C for 6 weeks.   Except for the killed control curves, each point represents the average of triplicate microcosms. Killed controls represent the average of 10 microcosms.

TABLE 1. Stoichiometry of naphthalene and methanol biodegradation under aerobic and denitrifying conditions.[a]

$$C_{10}H_8 + 5.83\ O_2 + 0.88\ NO_3^- + 0.88\ H^+ \Rightarrow 0.88\ C_5H_7O_2N + 5.6\ CO_2 + 1.34\ H_2O$$

$$C_{10}H_8 + 5.66\ NO_3^- + 5.66\ H^+ \Rightarrow 0.86\ C_5H_7O_2N + 5.71\ CO_2 + 2.40\ N_2 + 3.81\ H_2O$$

$$CH_3OH + 0.49\ O_2 + 0.14\ NO_3^- + 0.14\ H^+ \Rightarrow 0.14\ C_5H_7O_2N + 0.29\ CO_2 + 1.57\ H_2O$$

$$CH_3OH + 0.62\ NO_3^- + 0.62\ H^+ \Rightarrow 0.12\ C_5H_7O_2N + 0.38\ CO_2 + 0.25\ N_2 + 1.87\ H_2O$$

(a) Equations were derived using McCarty's (1971) model recognizing cell synthesis during degradation.

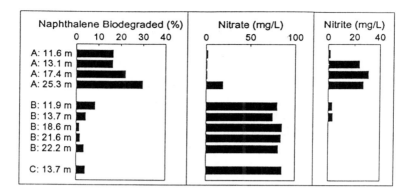

FIGURE 2. The ^{14}C-naphthalene mineralization, nitrate, and nitrite concentrations measured in nitrate-amended microcosms after 6 weeks of incubation. The Y-axis labels indicate the Borehole (A, B, or C) and sample depth.

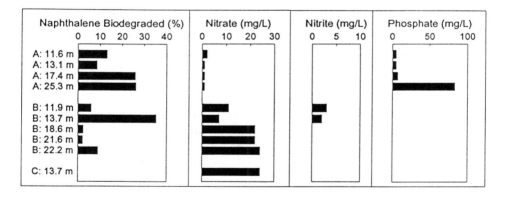

FIGURE 3. The ^{14}C-naphthalene mineralization, nitrite, nitrate, and phosphate concentrations measured in microcosms amended with nitrate and phosphate after 6 weeks of incubation. The Y-axis labels indicate the Borehole (A, B, or C) and sample depth.

These data suggest a predominance of facultative denitrifying bacteria in the Borehole A, 11.6-m and 13.1-m sediments. Stoichiometric relationships indicate that the amount of NO_3^- consumed in the microcosms initially receiving 85 mg/L NO_3^- far exceeded the amount necessary for naphthalene mineralization (Table 1). Since the fraction of natural organic carbon in the sediment is relatively low ($f_{oc} < 0.001$) it is likely that methanol served as an electron donor in denitrification.

In four out of the six active sediments, NO_3^- addition failed to enhance mineralization. In the Borehole B 11.9-m and 13.7-m samples, NO_3^- (85 mg/L) actually inhibited mineralization relative to microcosms not receiving nutrient amendments. For this sample, the degree of inhibition was significantly greater in the microcosms amended with 85 mg/L NO_3^- than in microcosms amended with 25 mg/L NO_3^- and 90 mg/L PO_4^{3-}. Inhibition by nitrate at comparable concentrations has been reported elsewhere and is thought to be related to either the oligotrophic nature of aquifer bacteria or shifts in the predominating populations and metabolic pathways of the consortia (Morgan and Watkinson 1992).

In most cases, naphthalene mineralization was either inhibited or unaffected by PO_4^{3-} amendments. The microcosms that had received 25 mg/L NO_3^- and 90 mg/L PO_4^{3-} generally did not behave differently from microcosms receiving 90 mg/L PO_4^{3-} without NO_3^-. Analysis of the supernatant at the end of the 6-week incubation indicates that, with the exception of one sample (Borehole A, 25.3 m), PO_4^{3-} was largely complexed to cations in the sediment (Figure 3). Thus, it is likely that PO_4^{3-} did not exert a significant effect for some of the sediments because of poor P bioavailability. If the PO_4^{3-} amendment concentrations had been above the sediment PO_4^{3-} complexation capacity, effects on mineralization may have been more apparent. In the Borehole B, 13.7-m sample, however, no aqueous PO_4^{3-} was observed in the supernatant, but microcosms amended with PO_4^{3-} were clearly able to mineralize naphthalene faster than unamended microcosms. Others have observed PO_4^{3-} to both enhance (Hutchins 1991; Efroymson and Alexander 1994) and inhibit (Swindoll et al. 1988; Morgan and Watkinson 1992) mineralization of aromatic compounds in soils.

Naphthalene mineralization in the absence of amendments was substantial (15 to 25%). In the 11.9-m sample from Borehole B, unamended microcosms biodegraded more naphthalene than those amended with nutrients. For some sediments at the BGE site, NO_3^- and PO_4^{3-} addition might not enhance bioremediation, and an adequate supply of nutrients might already be available in the sediment. Biodegradation in the absence of added N and P is often feasible in subsurface sediments because of nutrient cycling (Alexander 1994).

Four of the ten sediments did not exhibit significant naphthalene mineralization. The water quality of inhibited microcosms was characterized by low pH and high SO_4^{2-} concentrations (Figure 4). The pH of the supernatant was less than 3.5 in three of the four sediments not showing naphthalene mineralization, and SO_4^{2-} concentrations in these sediments were between 600 and 1,300 mg/L. The supernatant in microcosms exhibiting significant naphthalene mineralization was less inhibitory, yielding pH values between 4.5 and 6.4, and SO_4^{2-} concentrations less than 110 mg/L.

CONCLUSIONS

This work highlights the variability in nutritional and electron-acceptor requirements for the biodegradation of arene compounds in sediments. Data

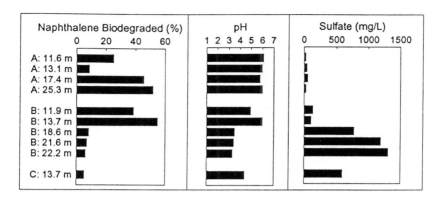

FIGURE 4. The ^{14}C-naphthalene mineralization, pH, and sulfate concentrations measured in high-oxygen (21 mg/L at start of experiment) microcosms after 6 weeks of incubation. The Y-axis labels indicate the borehole (A, B, or C) and sample depth.

show that subsurface bacteria from a former MGP site are capable of biodegrading naphthalene under aerobic, low-nutrient, and cold-temperature (10°C) conditions. Naphthalene mineralization could be enhanced in most sediments with high initial concentrations of oxygen (21 mg/L). Supplements of NO_3^- and/or PO_4^{3-} either enhanced, had no effect, or inhibited naphthalene mineralization. NO_3^- supplements (85 mg/L) enhanced mineralization more effectively than O_2 supplements in some samples, but inhibited mineralization in others. Use of a carrier solvent (methanol) to deliver ^{14}C-naphthalene confounded efforts to determine whether NO_3^- was used as an electron acceptor in the biodegradation of naphthalene. Finally, the finding that PO_4^{3-} inhibited mineralization in some sediments was unexpected. Further research is necessary to elucidate the mechanisms by which PO_4^{3-} can affect aquifer bacteria.

ACKNOWLEDGMENTS

This research was made possible through the generous support of the Baltimore Gas & Electric Company. We are grateful to EA Engineering and Science Inc. for assistance in gathering subsurface sediment and groundwater samples. We thank P. C. D'Adamo for a helpful review of the manuscript.

REFERENCES

Alexander, M. 1994. *Biodegradation and Bioremediation*. Academic Press, Inc., San Diego, CA. pp. 196–221.

Bone, T.L., and D.L. Balkwill. 1988. "Morphological and cultural comparison of microorganisms in surface soil and subsurface sediments at a pristine study site in Oklahoma." *Microbial Ecology* 16: 49-64.

Durant , N.D., L.P. Wilson, E.J. Bouwer. 1995. "Microcosm studies of subsurface PAH-degrading bacteria from a former manufactured gas plant." *Journal of Contaminant Hydrology* 17: 213-237.

Efroymson, R.A., and M. Alexander. 1994. "Biodegradation in soil of hydrophobic pollutants in nonaqueous-phase liquids (NAPLs)." *Environmental Toxicology and Chemistry* 13: 405-411.

Herbes, S.E., and L.R. Schwall. 1978. " Microbial transformation of polycyclic aromatic hydrocarbons in pristine and petroleum-contaminated sediments." *Applied and Environmental Microbiology* 35: 306-316.

Hutchins, S.R. 1991. "Optimizing BTEX biodegradation under denitrifying conditions." *Environmental Toxicology and Chemistry* 10: 1437-1448.

Madsen, E.L., J.L. Sinclair, and W.C. Ghiorse. 1991. "In situ biodegradation: microbiological patterns in a contaminated aquifer." *Science* 252: 830-833.

McCarty, P.L. 1971. "Energetics of bacterial growth." In *Organic Compounds in Aquatic Environments*. Marcel Dekker, Inc., New York, NY.

Morgan, P., and R.J. Watkinson. 1992. "Factors limiting the supply and efficiency of nutrient and oxygen supplements for the in situ biotreatment of contaminated soil and groundwater." *Water Research* 26: 73-78.

Park, K.S., R.C. Sims, and R.R. Dupont. 1990. "Transformation of PAHs in Soil Systems." *Journal of Environmental Engineering* 116: 632-640.

Robertson, B.K., and M. Alexander. 1992. "Influence of calcium, iron, and pH on phosphate availability for microbial mineralization of organic chemicals." *Applied and Environmental Microbiology* 58: 38-41.

Swindoll, M.C., C.M. Aelion, and F.K. Pfaender. 1988. "Influence of inorganic and organic nutrients on aerobic biodegradation and on the adaptation response of subsurface microbial communities." *Applied and Environmental Microbiology* 54 : 212-217.

Intrinsic Dechlorination of Trichloroethene to Ethene in a Bedrock Aquifer

David Major, Evan Cox,
Elizabeth Edwards, and Paul Hare

ABSTRACT

Trichloroethene (TCE) is being sequentially dechlorinated to ethene in a fractured bedrock aquifer system beneath and downgradient from evaporation pits and a closed waste-solvent storage tank at an inactive manufacturing facility in upstate New York. The relative lateral distributions of TCE, 1,2,-dichloroethene (1,2-DCE), vinyl chloride (VC), and ethene in the groundwater in the bedrock indicate that the migration of the chlorinated ethenes is being impacted by biologically mediated dechlorination. The biological activity has developed and is being maintained by acetone and methanol in the groundwater, which were disposed of along with TCE in the evaporation pits and waste-solvent storage tank. Evidence of sulfate depletion and methane and acetate production indicates that sulfate-reducing, methanogenic, and acetogenic bacteria are active in the bedrock. Isotopic analysis of methane indicates that the origin of the methane is biotic and not thermogenic. The distribution of the carbon and deuterium isotopes of the methane indicates that the methane was produced in the bedrock. This study provides evidence that microorganisms in a fractured bedrock aquifer can intrinsically dechlorinate TCE to ethene.

INTRODUCTION

Field and laboratory studies have shown that chlorinated volatile organic compounds (VOCs) can be can be biodegraded to nontoxic end products under anaerobic conditions. For example, Major et al. (1991), Fiorenza et al. (1994), and Wilson et al. (1994) have shown that chlorinated VOCs such as tetra-chloroethene and TCE can be intrinsically biodegraded to ethene by indigenous methanogenic, acetogenic, and sulfate-reducing bacteria.

At an inactive manufacturing facility located in upstate New York, TCE and its dechlorination products, 1,2-DCE and VC, have been detected in a fractured bedrock aquifer system downgradient from former evaporation pits and a closed waste-solvent storage tank. Acetone and methanol have also been detected in the fractured bedrock groundwater. An examination of the VOC-distribution data suggested that biodegradation of the chlorinated VOCs was limiting the migration of selected VOCs in the groundwater. Based on the site data and previous research, we conducted this study to determine whether the dechlorination processes are proceeding to completion in the fractured bedrock (i.e., chlorinated VOCs are being biodegraded to innocuous nonchlorinated end products such as ethene and ethane), and to evaluate the types of microbial processes that are associated with the biodegradation (i.e., methanogenesis, acetogenesis, and sulfate reduction).

Site Description

The site is underlain by 3 to 7 m of overburden that is generally fine-grained and of low permeability. A thick sequence of carbonate strata, consisting of approximately 14 m of limestone (shallow bedrock) and 37 m of dolomites (deep bedrock), underlies the overburden. The water table is located in the overburden just below ground surface in the late fall, winter, and early spring; however, significant desaturation of the overburden occurs during the summer as a result of evapotranspiration. A temporally persistent water-table divide occurs immediately to the east of the plant building and is oriented in a north-south direction. The former evaporation pits and waste-solvent tank are located west of the divide, and groundwater flow in these areas is generally to the northwest (see Figure 1a). Horizontal flow velocities of 0.18 and 1.6 m per day have been estimated for the overburden and shallow bedrock, respectively. Flow in the shallow bedrock is believed to be primarily along bedding planes that may be widened somewhat by solution. The groundwater at the site is predominantly anaerobic, characterized by negative oxidation-reduction potential and dissolved oxygen depletion.

METHODS

Groundwater samples were collected from 20 monitoring wells located along the groundwater flowpath upgradient (background), transgradient, and downgradient from the former evaporation pits and waste-solvent tank. Of the monitoring wells sampled, 2 were screened in the overburden, 13 were screened in the shallow bedrock, and 5 were screened in the deep bedrock. The lateral distribution of chlorinated VOCs in the groundwater is greater in the shallow bedrock than in the overburden or deep bedrock; therefore, the shallow bedrock was the focus of this study.

Groundwater samples for analysis of VOCs, dissolved hydrocarbon gases (ethene, ethane, and methane), oxygenates (methanol, acetone), acetic acid, methane isotopes, and sulfate were collected using dedicated Waterra™ pumps. Samples for analysis of volatile parameters were filled without headspace.

Samples were appropriately preserved, placed on ice, and delivered or shipped by express courier to the analytical laboratories for analysis by the methods described below.

VOCs (including acetone) were analyzed by gas chromatography/mass spectrometry (GC/MS) using United States Environmental Protection Agency (U.S. EPA) Method 8240. Sulfate was analyzed by ion chromatography using standard U.S. EPA methods. Dissolved hydrocarbon gases were analyzed by gas-phase injection/gas chromatography using a Hewlett Packard 5840A gas chromatograph with a flame ionization detector (FID). Oxygenates were analyzed by aqueous injection/gas chromatography using a Hewlett Packard 5840A gas chromatograph with a FID detector. Acetic acid was analyzed by ion chromatography using a Waters Ion Exclusion column with a conductivity detector. Carbon and deuterium isotopes of methane were analyzed by GC/MS.

RESULTS AND DISCUSSION

Distribution of Chlorinated VOCs, Ethene, and Ethane

The presence of 1,2-DCE and VC in the groundwater provides evidence that TCE is being biodegraded at the site. These degradation products were not used or produced at the site, and therefore their presence can only be attributed to the dechlorination of TCE. Ethene was detected in the groundwater samples, at concentrations ranging up to 17,110 μg/L, providing evidence that the chlorinated ethenes are being biodegraded to nontoxic end products. Ethane was not detected in the groundwater samples.

The lateral distributions of TCE, 1,2-DCE, VC, and ethene in the shallow bedrock groundwater, shown in Figures 1a to 1d, indicate that the migration of the chlorinated VOCs in the groundwater is being impacted by biodegradation. The lateral distribution of TCE is much smaller than the lateral distributions of 1,2-DCE and VC, and approximately corresponds to the lateral distribution of key nutrients such as acetone and acetic acid (see below) in the vicinity of the suspected source areas. The lateral distribution of the higher concentrations of 1,2-DCE and VC (greater than 100 μg/L) also mirrors the lateral distribution of the key nutrients, suggesting that a biologically active zone (BAZ) has formed around the suspected source areas at the site.

Distribution of Oxygenates

The lateral distribution of acetone in the groundwater is small relative to the lateral distributions of 1,2-DCE, VC, and ethene, and appears to be confined to an area surrounding the suspected source areas at the site (Figure 2a). Methanol was only reported in one well, located in the immediate vicinity of one of the former evaporation pits. The limited lateral distributions of acetone and methanol suggest that they are being biodegraded in the groundwater.

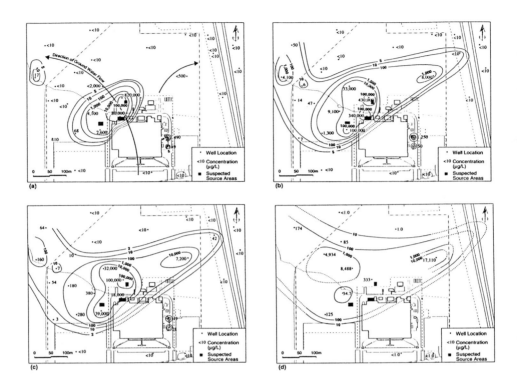

FIGURE 1. Isoconcentration contours of (a) trichloroethene, (b) 1,2-dichloroethene, (c) vinyl chloride, and (d) ethene in the shallow bedrock groundwater (contours dashed where inferred).

The literature indicates that methanol is an effective substrate for enhancing anaerobic dechlorinating activity (Freedman and Gossett 1989; Major et al. 1991). Methanol can be readily metabolized by acetogenic bacteria to acetate and hydrogen, which can be metabolized by sulfate-reducing and methanogenic bacteria (DiStefano et al. 1992). Acetone is also a very effective growth substrate for anaerobic bacteria.

Distribution of Methane, Acetic Acid, and Sulfate

The distribution of acetic acid, sulfate, and methane was used as an indirect measure of the activity of functional groups of microorganisms in the bedrock. The lateral distributions of acetic acid, sulfate, and methane in the shallow bedrock groundwater are shown in Figures 2b to 2d, respectively. The presence of acetic acid and methane, and the depletion of sulfate provide indirect evidence that acetogenic, methanogenic, and sulfate-reducing bacteria are active in the shallow bedrock at the site. The activity of each of these groups of microorganisms has been associated with the dechlorination of chlorinated

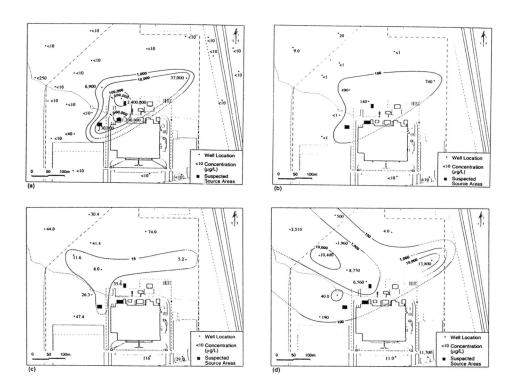

FIGURE 2. Isoconcentration contours of (a) acetone, (b) acetic acid, (c) sulfate, and (d) methane in the shallow bedrock groundwater (contours dashed where inferred).

ethenes.   In the presence of available sulfate, sulfate reduction tends to suppress methanogenesis because sulfate-reducing bacteria have a greater affinity for hydrogen and acetate and outcompete methanogenic bacteria (Gossett et al. 1992).   However, these microorganisms can coexist in the same habitat in the presence of a complex web of cofactors, end products, and inter-species hydrogen-transfer pathways that promote the establishment of a healthy dechlorinating microbial community.

Distribution of Methane Isotopes

Measurement of the carbon isotope ($^{13}C/^{12}C$) and hydrogen isotope ($^{1}H/^{2}H$) ratios of methane was conducted to separate biogenic and thermogenic (abiotic) methane, and to indicate where biological activity may have been occurring. Thermogenic methane deposits have been reported in the vicinity of the site. Microorganisms preferentially use the lighter isotopes of carbon and hydrogen and, therefore, methane derived from biochemical reactions is isotopically lighter than thermogenic methane.

The carbon and hydrogen isotope ratios of methane in groundwater samples collected from wells located within the proposed biologically active zone (BAZ) were consistently enriched with the lighter isotopes compared to background samples, indicating that the methane was produced biologically. More importantly, the data indicate that the methane was produced in the shallow bedrock, and is not present in the shallow bedrock as a result of downward transport from the overburden unit. At a well nest located within the BAZ, the methane in the shallow bedrock groundwater was more enriched in the lighter isotopes than the overburden groundwater at the same location. If the methane had been produced in the overburden and transported to the shallow bedrock, methane isotope dilution would have been observed instead.

CONCLUSIONS

The lateral distributions of acetone, methanol, acetic acid, methane, sulfate and the chlorinated VOCs suggests that a BAZ has formed around the suspected source areas at the site. An active and diverse anaerobic microbial community, consisting of sulfate-reducing, methanogenic, and acetogenic bacteria has been established and appears to be maintained by acetone and methanol. TCE is being completely dechlorinated within the BAZ. 1,2-DCE and VC are being dechlorinated in the BAZ; however, lower concentrations of these VOCs (i.e., less than 100 μg/L) emanate from the downgradient side of the BAZ. The residence time within the BAZ may not be sufficient to allow complete dechlorination of 1,2-DCE and VC to ethene. Overall, this study provides evidence that microbial populations in a fractured bedrock aquifer can intrinsically dechlorinate TCE to ethene when suitable substrates are present to support their growth.

REFERENCES

DiStefano, T. D., J. M. Gossett, and S. H. Zinder. 1992. "Hydrogen as an Electron Donor for Dechlorination of Tetrachloroethene by an Anaerobic Mixed Culture." *Appl. Environ. Microbiol. 58*: 3622-3629.

Fiorenza, S., E. L. Hockman Jr., S. Szojka, R. M. Woeller, and J. W. Wigger. 1994. "Natural Anaerobic Degradation of Chlorinated Solvents at a Canadian Manufacturing Plant." In R. E. Hinchee, A. Leeson, L. Semprini, and S. K. Ong (Eds.), *Bioremediation of Chlorinated and Polycyclic Aromatic Hydrocarbon Compounds,* Lewis Publishers, Boca Raton, FL, pp. 277-286.

Freedman, D. L., and J. M. Gossett. 1989. "Biological Reductive Dechlorination of Tetrachloroethylene and Trichloroethylene to Ethylene Under Methanogenic Conditions." *Appl. Environ. Microbiol. 55*: 2144-2151.

Gossett, J. M., T. D. DiStefano, and S. H. Zinder. 1992. "The Role of Hydrogen in the Biotransformation of Chlorinated Ethenes by an Anaerobic Mixed Culture: Implications for Bioremediation." In S. Lesage (Ed), *In Situ Bioremediation Symposium '92.* Environment Canada, Niagara-on-the-Lake, Ontario, Canada.

Major, D. W., E. H. Hodgins, and B. H. Butler. 1991. "Field and Laboratory Evidence of In Situ Biotransformation of Tetrachloroethene to Ethene and Ethane at a Chemical Transfer Facility in North Toronto." In R. E. Hinchee and R. F. Olfenbuttel (Eds.), *In Situ and On Site Bioreclamation,* Butterworth-Heinemann, Stoneham, MA.

Wilson, J. T., J. W. Weaver, and D. H. Kampbell. 1994. "Intrinsic Bioremediation of TCE in Ground Water at an NPL site in St. Joseph, Michigan." *Symposium on Intrinsic Bioremediation of Groundwater.* U.S. Environmental Protection Agency, EPA/540/R-94/515, August 1994, Denver, CO.

Intrinsic In Situ Anaerobic Biodegradation of Chlorinated Solvents at an Industrial Landfill

Michael D. Lee, Paul F. Mazierski,
Ronald J. Buchanan, Jr., David E. Ellis,
and Lily S. Sehayek

ABSTRACT

The DuPont Necco Park Landfill in Niagara Falls, New York, is contaminated with numerous chlorinated solvents at concentrations of up to hundreds of mg/L in the groundwater. An extensive monitoring program was conducted to determine if intrinsic anaerobic biodegradation was occurring at the site, to determine what might limit this activity, and to characterize this activity with depth and distance away from the landfill. It was determined that anaerobic microbial activity was occurring in all zones, based upon the presence of intermediate products of the breakdown of the chlorinated solvents and the presence of final metabolic end products such as ethene and ethane. Aerobic, iron-reducing, manganese-reducing, sulfate-reducing, and methanogenic redox conditions were identified at the site. High levels of nitrogen and biodegradable organic compounds were present in most areas to support cometabolic anaerobic microbial activity against the chlorinated solvents. Intrinsic biodegradation is clearly evident and is effective in reducing the concentrations of chlorinated organic in the groundwater at the site. Groundwater modeling efforts during development of a site conceptual model indicated that microbial degradation was necessary to account for the downgradient reduction of chlorinated volatile organic compounds as compared to chloride, a conservative indicator parameter.

INTRODUCTION

The objective of this research was to determine whether chlorinated solvents are being biodegraded at an industrial landfill in Niagara Falls, New York, which is owned by the DuPont company. Process wastes, including chlorinated solvents, barium, and other materials were placed in the Necco Park

Landfill over a period of 50 years. The landfill no longer receives waste materials. A cap has been placed over the landfill and a pump-and-treat system has been installed to contain leachate.

This site is representative of many landfills, where a large variety of industrial chemicals were disposed of and are contaminating the groundwater. Biodegradation processes occurring at this site and other landfills may reduce or contain the contamination problems. Intrinsic bioremediation is an acceptable remediation alternative for some sites contaminated with chlorinated solvents (Vogel 1994).

SITE CHARACTERISTICS

Figure 1 presents a map of the Necco Park Landfill. A commercial landfill containing residential, industrial, and hazardous wastes has been constructed on three sides of the Necco Park Landfill. Many wells have been installed at and around the site to delineate groundwater movement and the extent of contamination in nine distinct groundwater fracture zones.

A relatively thin, low-conductivity overburden designated as the A zone is the first geologic unit at the site (Figure 2). A fractured dolomitic bedrock of the Lockport Group underlies the A zone. The wastes were placed in the overburden and contaminants have leached into the deeper zones. The primary groundwater flow occurs primarily through nine horizontally continuous bedding plane fracture zones designated as the B through G zones. The J zone corresponds to the interface between the bottom of the Lockport Group and the Rochester Shale. Because groundwater flow through the A and J zones is

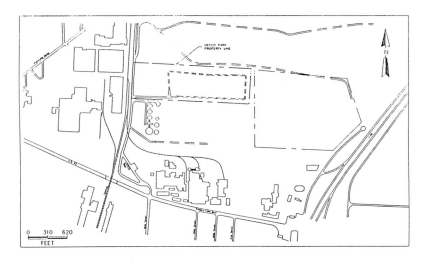

FIGURE 1. Necco Park site map.

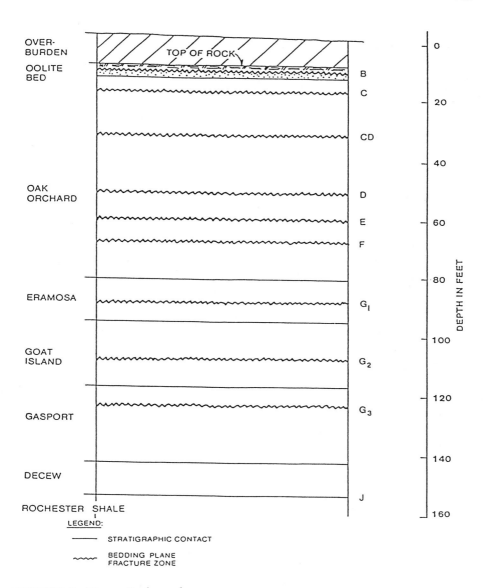

FIGURE 2. Necco Park geology.

very limited, these zones were not included in the transport modeling conducted to develop a conceptual model for the site.

The focus of this paper will be on selected wells in the B and F zones that are representative of the processes occurring in the other zones. The B zone is one of the most highly contaminated zones. There are lower levels of contaminants in the F zone, but much of the potential groundwater flow away from the site is in this zone.

GEOCHEMICAL EVALUATION

The geochemical evaluation was conducted to gain an understanding of the site's geochemistry and the potential for intrinsic bioremediation. Specific objectives of the geochemical evaluation were to:

- Determine if anaerobic reductive dehalogenation was naturally occurring at the site
- Determine what factors might control the rates of dehalogenation
- Characterize the groundwater chemistry for the potential implementation of an in situ bioremediation project at the site.

The contaminants include a number of chlorinated compounds, including chloroethenes, chloroethanes, chloromethanes, chlorophenols, chlorobenzenes, and other compounds. This paper will focus on the distribution and concentrations of the chloroethenes, chloroethanes, and chloromethanes and their daughter products because they were the major contaminants. The abbreviations used in this text are defined in Table 1. The degradation for the chloroethenes proceeds sequentially from perchloroethene (PCE) to trichloroethene (TCE) to *cis*-1,2-dichloroethene (*c*DCE) predominantly, with *trans*-1,2-dichloroethene *(t*DCE) and 1,1-dichloroethene (1,1-DCE) as minor products. The next degradation product is vinyl chloride (VC) with subsequent transformation to ethene. A similar degradation pattern of reductive dechlorination by the stepwise removal of chloride is followed by the chloroethanes and chloromethanes.

The wells in each zone were divided into the following classification based upon the concentrations of total volatile organic contaminants (TVOC):

- Background nondetect to 0.1 mg/L TVOC
- Slightly contaminated 0.1 to 10 mg/L TVOC
- Moderately contaminated 10 to 100 mg/L TVOC
- Highly contaminated >100 mg/L TVOC.

TABLE 1. List of organic contaminants found at Necco Park site.

Chloroethenes	Chloroethanes	Chloromethanes
Perchloroethene (PCE)	Hexachloroethane (HCA)	Carbon Tetrachloride (CT)
Trichloroethene (TCE)	Pentachloroethane (PCA)	Chloroform (CF)
cis-1,2-Dichloroethene (CDCE)	Tetrachloroethane (TeCA)	Dichloromethane (DCM)
trans-1,2-Dichloroethene (TDCE)	1,1,1-Trichloroethane (1,1,1-TCA)	Chloromethane (CM)
1,1-Dichloroethene (1,1-DCE)	1,1,2-Trichloroethane (1,1,2-TCA)	
Vinyl chloride (VC)	1,1-Dichloroethane (1,1-DCA)	
	1,2-Dichloroethane (1,2-DCA)	
	Chloroethane (CA)	

Table 2 provides a list of the geochemical evaluation analytes that were measured in the groundwater of the selected wells. Some wells were analyzed for a less extensive list of groundwater parameters, which are routinely measured at the site. The samples were collected in March and April, 1993. This paper will focus on the following parameters in selected wells in the B and F zones:

- Field parameters, including pH, temperature, redox potential, and dissolved oxygen
- Chloride as a nonreactive tracer
- Organic substrates that may serve as electron donors, measured by total organic carbon (TOC) and biochemical oxygen demand (BOD5)
- Inorganic nutrients (nitrogen and phosphorus) and electron acceptors needed to support growth
- Final metabolic end products including methane and sulfide
- Evidence for dechlorination of chloroethenes, chloroethanes, and chloromethanes, including the presence of daughter products, and completely reduced products, such as ethene and ethane.

The redox status for each well was assigned according to the groundwater geochemistry criteria developed by Lyngkilde et al. (1991) for a landfill in the Netherlands. The concentrations of the final electron acceptors in the series are the most important; if methane were to exceed 1.0 mg/L, then this portion of the aquifer would be considered methanogenic. The aquifer was under sulfate-reducing conditions if the methane was less than 1.0 mg/L and the sulfide was greater than 0.2 mg/L. Iron-reducing concentrations would predominate

TABLE 2. List of analytical parameters measured during geochemical evaluation.

Nutrients/ Electron Acceptors	Substrates	Field Parameters	Metabolic End Products	Other
Nitrate-Nitrogen (NO_3-N)	Total Organic Carbon (TOC)	pH	Methane	Chloride (Cl)
Total Kjeldahl Nitrogen (TKN)	Biochemical Oxygen Demand (BOD5)	Temperature (Temp.)	Ethene	Heterotrophic Bacteria (HET)
Ammonia-Nitrogen (NH_3-N)		Redox Potential (Eh)	Ethane	Sulfate-Reducing Bacteria (SRB)
Total Phosphorus (Tot. P) Sulfate (SO_4)		Dissolved Oxygen (DO)	Sulfide S^{2-}	
Soluble Iron (Sol. Fe)				

if the Fe^{2+} levels were >1.5 mg/L, sulfide <0.1 mg/L, and methane <1.0 mg/L. Manganese-reducing conditions would exist where Mn^{2+} was >0.2 mg/L, iron <1.5 mg/L, sulfide <0.1 mg/L, and methane <1.0 mg/L. Nitrate-reducing conditions would exist when nitrate and nitrite were present, but the other electron acceptors were not present above the critical levels. Oxygen is the preferred electron acceptor at dissolved oxygen (DO) levels above 1.0 mg/L.

GROUNDWATER MODELING

Groundwater modeling was conducted during preparation of a site-specific conceptual model. The U.S. Environmental Protection Agency (EPA) Multimedia Exposure Assessment Model (MULTIMED) developed by Salhotra et al. (1990) was used to model chemical transport in distinct fractures of the Lockport Formation. The model allowed an estimate to be made of the relative importance of biodegradation. The following assumptions were made:

- No interactions between the zones outside the source area
- Flow in each zone is uniform and one-dimensional
- Continuous patch source with constant strength.

The model simulated the dilution of chloride and total dissolved volatile organic contaminants along the radial distance between wells within a zone. The transport model was run on a regional scale. The model considers dilution, dispersion, and adsorption. Other removal mechanisms can be incorporated as first-order degradation rates. The results of the model for the F zone are presented in this paper. The observed chloride dilution factors were used in calibration and verification of hydrogeologic parameters. The observed TVOC dilution factors were used in demonstrating the importance of biodegradation. The fate of the individual organic contaminants was not addressed in this simplified computer simulation.

B ZONE

Figure 3 presents the map of the isocontours for TVOC in the wells that monitor the B zone. Groundwater flow is from north to south. The geochemical evaluation focuses on wells 142B and 148B in the background area with between nondetect and 0.1 mg/L TVOC. The only well sampled for the complete geochemical evaluation list in the slightly contaminated area (with 0.1 to 10 mg/L TVOC) is 137B. Wells 112B and 130B in the moderately contaminated areas (with 10 to 100 mg/L TVOC) are discussed, as are highly contaminated-area wells 111B and 129B (with greater than 100 mg/L TVOC).

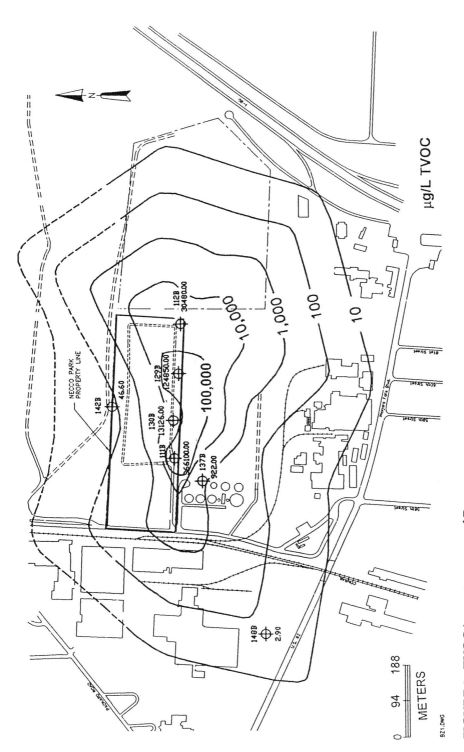

FIGURE 3. TVOC isocontour map of B zone.

B-Zone Nutrients, Substrates, Field Parameters, and Electron Acceptors

The concentrations of the electron acceptors, field parameters, natural organic substrates, and nutrients in the selected wells in the B zone are given in Table 3. The background wells contained 1.6 and 10 mg/L DO, and redox potentials (Eh) of 94 and 300 mV. Well 148B was under aerobic conditions. The measured DO levels ranged between 0.8 to 2.4 mg/L in the more contaminated wells. The DO was higher than expected for the contaminated wells that had elevated substrate levels and low redox potentials, suggesting that oxygen was introduced into the wells during sampling. The Eh was generally lower in the contaminated wells, with two wells showing negative redox potentials. There were elevated pHs of greater than 9.0 in a number of the wells, primarily as a result of placement of high-H wastes into the Necco Park Landfill. Background well 142B had little organic contamination, but showed a history of exposure to the materials placed into the landfill based upon the high pH and elevated chlorides, ammonia, total Kjeldahl nitrogen (TKN), substrates, and sulfide. The temperature throughout the B zone ranged between 6.1 and 10.5°C during the March to April sampling period. Chloride levels generally increased with the organic contamination concentrations. There was little organic substrate in background well 148B, as evidenced by the low TOC and BOD5 and the elevated DO. In general, there were higher levels of organic substrates to support biodegradation of the chlorinated solvents in the more highly contaminated wells.

TABLE 3. Electron acceptors, field parameters, and nutrients for B-zone wells.

Classification TVOC Conc.	Background ND to 0.1 mg/L		Slightly 0.1 to 10 mg/L	Moderately 10 to 100 mg/L		Highly >100 mg/L	
Well	142B	148B	137B	112B	130B	111B	129B
DO (mg/L)	1.6	10	1.2	1.1	1.9	2.4	0.8
Eh (mV)	94	300	-52	32	-53	120	83
pH	12.6	7.7	13	10.5	8.0	9.6	8.0
Temp. (°C)	6.5	6.5	7.0	10.5	9.0	9.5	6.1
Cl (mg/L)	150	20	1,400	1,900	3,000	8,800	29,000
TOC (mg/L)	46	3.7	99	130	88	1,900	210
BOD5 (mg/L)	54	<2	100	160	150	2,200	200
NH_3-N (mg/L)	16	<0.5	26	110	90	240	110
NO_3-N (mg/L)	<0.5	0.6	<0.5	<0.5	<0.5	<0.5	<0.5
TKN (mg/L)	20	6.4	31	79	170	430	120
Tot. P (mg/L)	0.12	0.14	<0.05	0.46	0.12	0.24	<0.05
Sol. Fe (mg/L)	0.07	0.06	0.11	0.09	1.8	28	130
SO_4 (mg/L)	23	62	18	210	540	<5	28
Sulfide (mg/L)	6.3	<1	1.2	33	10	1.8	6.6
Methane (mg/L)	3.3	<0.001	9.2	8.0	3.4	2.2	2.1

Ammonia was found at high concentrations in many of the wells. Nitrogen availability should not limit biodegradation in this zone. There was little nitrate. Considerable organic nitrogen was present based upon the TKN analyses, which include ammonia and organic nitrogen. Very little phosphorus was found in any of the B-zone wells. Iron was being used as an electron acceptor in wells 130B, 111B, and 129B based upon soluble iron concentrations greater than 1 mg/L. Plate counts of anaerobic heterotrophs ranged between $<1 \times 10^1$ and 3×10^2 colony-forming units per mL (cfu/mL); these plate counts are likely to represent only a fraction of the total microbial population. Sulfate-reducing bacteria (SRB) were also present, based on the high levels of sulfide in all of these wells, except 148B, and on plate counts made on a selective medium to isolate sulfate reducers. Wells 112B and 130B had SRB counts of up to 8×10^4 cfu/mL, but SRBs were not isolated from the other wells in the B zone. The availability of sulfate could limit the activity of sulfate-reducers in the highly contaminated wells. Because methane concentrations were greater than 1.0 mg/L in all of the B-zone wells (with the exception of 148B), and because there were low sulfate levels (less than 30 mg/L in wells 142B, 137B, 111B, and 129B), the final electron acceptor in most of the B zone wells was probably carbon dioxide. However, dehalogenation of chloromethanes can also generate methane, so there is no conclusive proof that methanogens were present.

B-Zone Chlorinated Organic Compounds

Table 4 presents the micromolar (μM) concentrations of the various chloroethenes in the selected B-zone wells. The μM concentrations are presented so that the concentrations of parent and daughter products are comparable. In the background area, there appeared to be some conversion of PCE to cDCE and ethene. In the slightly contaminated area well 137B, the complete sequence of PCE to ethene was observed, with the largest accumulations of TCE, cDCE, VC, and ethene. The moderately contaminated area wells also showed the complete degradation sequence from PCE to ethene. Well 111B has an unusual pattern, with higher concentrations of 1,1-DCE than cDCE. Well 111B may have received 1,1-DCE, or a somewhat unusual transformation may have occurred in this area. The other highly contaminated area well, 129B, had the complete sequence of chlorinated ethenes with a predominance of TCE and cDCE.

The chloroethanes were generally present at lower concentrations than the chloroethenes. Chloroethane was never detected in this study, either because it is rapidly degraded or because the analytical method is less sensitive for this compound. Only dichloroethane (DCA) was found in the slightly contaminated well 137B. Ethane was detected in 112B. The degradation sequence from tetrachloroethane (TeCA) to trichloroethane (TCA) to DCA was observed in 130B, with an accumulation of TCA. In the highly contaminated wells, the degradation sequence from hexachloroethane (HCA) to ethane occurred in 111B (with the exception of TeCA), and from TeCA to ethane in 129B.

TABLE 4. Concentrations (μM) of chloroethenes, chloroethanes, and chloro-methanes in B-zone wells.

Classification TVOC Conc. Well	Background ND to 0.1 mg/L		Slightly 0.1 to 10 mg/L	Moderately 10 to 100 mg/L		Highly >100 mg/L	
	142B	148B	137B	112B	130B	111B	129B
PCE	0.19		0.54	15	3.4	160	15
TCE	0.084	0.009	2.6	140	14	1,400	270
cDCE	0.047	0.018	2.2	5.0	44	59	140
tDCE			0.29		3.4		29
1,1-DCE			0.42		5.6	210	8.9
VC			2.7		54		48
Ethene	1.9		33	8.9	43	50	14
HCA						0.55	
TeCA					3.4		77
1,1,2-TCA					6.4	230	110
1,2-DCA			0.14		0.57	43	7.0
Ethane				2.3		5.3	2.7
CT						59	
CF			0.25	80	5.9	670	330
DCM		0.046	1.3	19	2.6	720	65

The degradation sequence of chloromethanes was observed in several of the wells. Carbon tetrachloride (CT), chloroform (CF), and dichloromethane (DCM) were observed in 111B. Chloromethane was never detected. Methane was detected in most of the wells. It may have been formed from the dehalogenation of the chloromethanes or the activity of methanogens.

F ZONE

Groundwater flow in the F zone is from east to west (Figure 4). The F zone was analyzed for the Necco Park Indicator List, which includes fewer parameters than the complete geochemical evaluation list used in the B zone. There were no wells in the F zone with greater than 100 mg/L TVOC. The F-zone wells included wells 141F and 153F/G in the background area, wells 136F and 156F in the slightly contaminated area, and wells 112F, 146F, 147F, and 156F in the moderately contaminated area.

F-Zone Nutrients, Substrates, Field Parameters, and Electron Acceptors

Table 5 presents the concentrations of nutrients, organic substrates, field parameters, and electron acceptors for the F zone. Wells 141F and 136F were alkaline, with pHs greater than 9.0. The temperature in the F zone ranged from

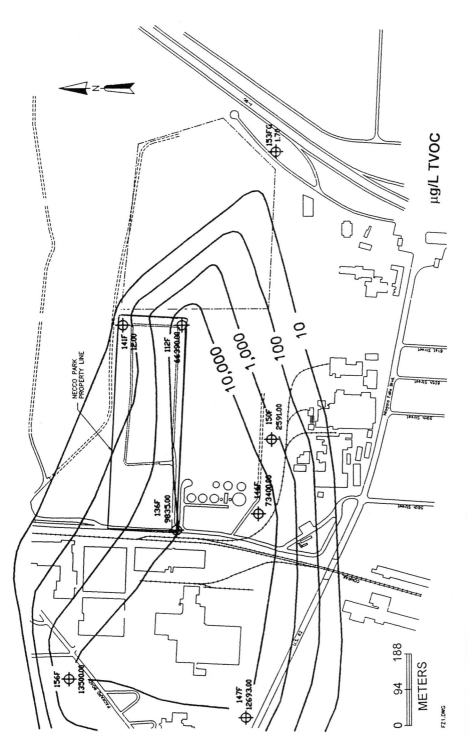

FIGURE 4. TVOC isocontour map of F zone.

TABLE 5. Electron acceptors, field parameters, and nutrients for F-zone wells.

Classification TVOC Conc.	Background ND to 0.1 mg/L		Slightly 0.1 to 10 mg/L			Moderately 10 to 100 mg/L		
Well	141F	153F/G	136F	156F	112F	146F	147F	156F
pH	9.7	7.5	9.2	7.5	7.3	7.8	7.4	7.7
Temp. (°C)	12	8.5	11	9.5	20	14	8.2	8.1
NH_3-N (mg/L)	51	5.4	16	120	5	51	8.0	39
Cl (mg/L)	370	6,100	490	7,600	570	1,400	320	850
TOC (mg/L)	9.9	<1	23	270	7.4	120	21	39
BOD5 (mg/L)	35	12	22	490	5.3	190	28	38

8.1 to 20°C, which was somewhat higher than in the shallower zones. Ammonia-nitrogen levels ranged between 5 and 120 mg/L. Chloride levels were somewhat variable and did not correlate consistently with increasing organic contaminant concentrations. Organic substrates were limited in the F zone, with only wells 156F and 146F containing more than 100 mg/L of TOC and BOD5.

F-Zone Chlorinated Organic Compounds

As shown in Table 6, only TCE and ethene were found in the background wells. The complete sequence of PCE to ethene was observed in wells 136F, 146F, and 156F. Wells 150F and 147F contained the complete sequence except for ethene. The majority of the chloroethenes in slightly and moderately contaminated area wells were found as TCE and cDCE. Well 156F had a very high concentration of VC (40 μM).

In general, there was less extensive transformation of the chloroethanes and chloromethanes found in the F zone than in the B zone. TeCA and TCA were the only chloroethane daughter products found in any F-zone well, except 156F, where DCA was detected (Table 6). The final degradation product, ethane, was found in only two wells; the background well 153F/G and the moderately contaminated 112F. The ethane detected in well 153F/G may have been carried with the groundwater from a more contaminated zone or may reflect prior metabolic activity against the chloroethanes in this area. DCM was not measured in the F zone. Only wells 112F and 147F contained both CT and CF, which suggested limited transformation of the chloromethanes in the F zone.

GROUNDWATER MODELING RESULTS

The following input parameters were used for the MULTIMED model:

- Porosity of 0.05

TABLE 6. Concentrations (μM) of chloroethenes, chloroethanes, and chloro-methanes in F-zone wells.

Classification TVOC Conc.	Background ND to 0.1 mg/L		Slightly 0.1 to 10 mg/L			Moderately 10 to 100 mg/L		
Well	141F	153F/G	136F	150F	112F	146F	147F	156F
PCE			2.2	0.31	17	27	2.1	2.4
TCE	0.091	0.013	29	1.1	62	220	45	21
cDCE			12	13	37	100	21	51
tDCE			0.89	1.1	9.2	22	9.1	7.0
1,1-DCE			1.0	1.1		10	3.2	2.0
VC			2.9	12		22	2.1	40
Ethene	2.5		1.6		1.6	3.5		9.3
HCA					5.5			
TeCA			1.3		250	14	1.6	1.1
TCA			11			47	9.8	9.8
DCA								1.8
Ethane		43			0.43			
CT					29		0.34	
CF			20	0.84	37	140	13	3.1

- Aquifer thickness of 1.83 m
- Hydraulic gradient of 0.0043
- Hydraulic conductivity of 25,600 m/yr. (8.1×10^{-1} cm/s)
- Source duration of 50 years
- Source width of 345 m
- Source length of 558 m.

The aquifer parameters were obtained from a site-specific numerical groundwater flow model, geologic studies conducted at Necco Park, and regional flow studies. The source duration was reasonable because the Necco Park site had been in use since the mid-1930s. The dimensions of the source area were based on site data. A sensitivity analysis indicated that the solution was not sensitive to realistic variations in the dimensions of the source. Dispersion values used in the simulation were a function of the distance traveled and were generated by the MULTIMED model.

The overall direction of groundwater movement is shown by the arrow based upon the piezometric map for the F zone (Figure 5). The groundwater flow direction between wells 146F and 156F and from wells 146F to 147F are not in line with the overall direction of groundwater flow. An angle away from the major direction of flow must be considered in the modeling to account for the

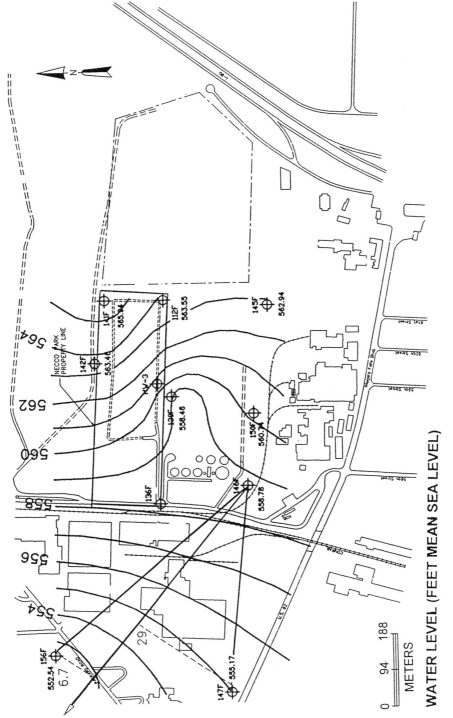

WATER LEVEL (FEET MEAN SEA LEVEL)

FIGURE 5. Groundwater flow direction in F zone.

difference in the dilution along the flow paths between wells 146F and 156F, and wells 146F and 147F, and the major direction of groundwater flow.

The output from the MULTIMED Model for the F zone is summarized in Table 7. The observed C/C_0 for chloride between wells 146F and 147F was 0.66. A C/C_0 for the TVOC was calculated to be 0.532. The model simulated a C/C_0 (i.e. contaminant concentration/initial concentration) for chloride of 0.7 using dispersion and dilution and assuming the observation well to be in the major direction of groundwater flow. Because the model did not consider the chloride generated during the dehalogenation of the chlorinated solvents, the dispersion and dilution predictions from the model represent a conservative estimate. With the angle between the plume center and the radial flow between the wells set at 6.7 degrees, a simulated C/C_0 of 0.65 was achieved. Using a decay rate of 0.7/y, the simulated C/C_0 for the TVOC was 0.53. This matched closely with the calculated C/C_0 of 0.532, which suggested that the first-order biodegradation coefficient of 0.7/y was appropriate. Retardation of the VOCs onto aquifer solids could not account for the observed VOC dilution factors. Unrealistically high values (i.e. greater than 100) of retardation of the VOCs were required to match the observed dilution factors.

The observed C/C_0 between wells 146F and 147F was 0.218, with a C/C_0 of 0.199 for TVOC. Without angle correction, the model suggested a C/C_0 for chloride of 0.79. Using an angle of 29 degrees counterclockwise from the direction of groundwater flow as a correction, a simulated C/C_0 of 0.22 was achieved. With the first-order decay rate of 0.7/y, the simulated C/C_0 for total volatiles was 0.19. This compares favorably with the calculated TVOC of 0.199.

The first-order biodegradation coefficient of 0.7/y shows that intrinsic biodegradation of the chlorinated solvents is occurring at the site and that much of the contamination will be biodegraded. The first-order biodegradation coefficient of 0.7/y is equivalent to a half-life of 1 year.

DISCUSSION

The intrinsic attenuation of chlorinated solvents has been reported at several sites. Major et al. (1991) documented the biotransformation of up to 27 μM

TABLE 7. Output of MULTIMED model for F zone.

Parameter	146F to 156F	146F to 147F
Observed C/C_0 Cl	0.66	0.218
Calculated C/C_0 TVOC	0.532	0.199
Simulated C/C_0 Cl	0.7	0.79
Angle between plume center and wells	6.7	29
Simulated C/C_0 Cl along radius of wells	0.65	0.22
Biodegradation coefficient	0.7/y	0.7/y
Simulated C/C_0 TVOC	0.53	0.19

tetrachloroethene to ethene and ethane at a chemical transfer facility in Canada, based on field evidence including overlying PCE, TCE, cDCE, VC, and ethene plumes. Organic solvents, such as methanol and other substrates in the groundwater, were thought to support the dechlorination reaction. Laboratory microcosm studies confirmed that sulfate-reducers, methanogens, and (potentially) acetogens were present in the soil and groundwater and were capable of transforming PCE to ethene.

Similarly, Fiorenza et al. (1994) observed intrinsic anaerobic degradation of chlorinated solvents at another Canadian site. Evidence for the intrinsic reductive dechlorination of maximum concentrations of 0.1 μM PCE, 11 μM TCE, 41 μM 1,1,1-TCA, and 440 μM DCM were described. Organic contaminants such as naphtha and volatile fatty acids may have served as electron donors at this site.

Wilson et al. (1994) documented the intrinsic bioremediation of up to 760 M TCE at a National Priority Site in St. Joseph, Michigan. They reported apparent TCE degradation constants of between –0.0076/week and –0.0024/week or the equivalent of –0.40 to –0.12/y Naturally occurring organic compounds, as measured by the chemical oxygen demand, were thought to support the dechlorination reaction (McCarty & Wilson 1992).

Contaminant concentrations were higher at the Necco Park site than at the sites reported by Major et al. (1991), Fiorenza et al. (1994), and Wilson et al. (1994). Maximum concentrations of 160 μM PCE, 1,400 μM TCE, 140 μM cDCE, 48 μM VC, 77 μM TeCA, 110 μM 1,1,2-TCA, 7.0 μM 1,2-DCA, 59 μM CT, 670 μM CF, and 720 μM DCM were observed at the Necco Park site. The organic substrates at the Necco Park site were not defined, but are likely to be related to the materials placed in the landfill, because the higher substrate levels were generally related to higher chlorinated solvent concentrations.

The fractured bedrock geology of the Necco Park site is also quite different from the geology of the North Toronto, Canada, site, which is a layered clay, silty/clay, and sand/silt aquifer (Major et al. 1991), and the sandy aquifer at St. Joseph, Michigan (McCarty & Wilson 1992). The Necco Park geology is similar to that of the site described by Fiorenza et al. (1994), which consisted of the following units with depth: fill containing reworked till, sands, and silts; a weathered, glacial till of either silty sand or sandy silt; and a fractured shale-limestone bedrock.

Modeling results for the F zone of the Necco Park site provided a TVOC half-life of 1.0 y, which is similar to the TCE half-lives of 1.7 to 5.8 y calculated for the St. Joseph, Michigan, site.

CONCLUSIONS

An extensive monitoring program at the Necco Park Landfill determined that intrinsic bioremediation of chlorinated solvents is occurring at the site. Anaerobic microbial activity occurs in all zones, based upon the presence of intermediate products of the breakdown of chlorinated solvents (such as cDCE,

VC, DCA, and DCM), and the presence of final metabolic end products (such as ethene and ethane). Anaerobic biodegradation of the chlorinated solvents was evident even in areas with high pH, high barium concentrations, high salinity, and high concentrations of the chlorinated solvents. Aerobic, iron-reducing, manganese-reducing, sulfate-reducing, and methanogenic redox conditions were identified at the site based upon the detection of oxygen, soluble iron, soluble manganese, sulfide, and methane in the groundwater.

Sufficient nitrogen and biodegradable organics were present in most areas to support microbial activity against the chlorinated solvents. Ammonia-nitrogen levels were in excess of 20 mg/L in the contaminated wells in the B zone and ranged from 5 to 120 mg/L in the F-zone wells with >0.1 mg/L TVOC. The $BOD5$ found in the B-zone contaminated wells ranged from 100 to 2,200 mg/L. Somewhat lower quantities of substrates were found in the TVOC-contaminated F-zone wells, with 5 to 490 mg/L BOD5. The low concentrations of phosphorus (< 1 mg/L), the low groundwater temperature of 9°C during the winter, the availability of electron acceptors, and the availability of organic substrates in the deeper zones may control microbial activity.

Groundwater modeling efforts during development of a site conceptual model indicate that microbial degradation was necessary to account for the downgradient reduction of chlorinated volatile organic compounds as compared to chloride, a conservative indicator parameter. An apparent TVOC first-order biodegradation rate coefficient of 0.7/y was estimated, based upon the modeling studies, and is sufficient to degrade much of the organic contamination on site. Intrinsic biodegradation is clearly evident at the Necco Park site and is effective in reducing the concentrations of the chlorinated organic compounds in the groundwater of the site.

REFERENCES

Fiorenza, S., E. L. Hockman, Jr., S. Szojka, R. M. Woeller, and J. W. Wigger. 1994. "Natural Anaerobic Degradation of Chlorinated Solvents at a Canadian Manufacturing Plant." In R. E. Hinchee, A. Leeson, L. Semprini, and S. K Ong (Eds.), *Bioremediation of Chlorinated and Polycyclic Aromatic Hydrocarbon Compounds*, pp. 277-286. Lewis Publishers, Boca Raton, FL.

Lyngkilde, J., T. H. Christensen, B. Skov, and A. Foverskov. 1991. "Redox Zones Downgradient of a Landfill and Implications for Biodegradation of Organic Compounds." In R. E. Hinchee and R. F. Olfenbuttel (Eds.), *In Situ Bioreclamation Applications and Investigations for Hydrocarbon and Contaminated Site Remediation*, pp. 363-376. Butterworth-Heinemann, Stoneham, MA.

Major, D. W., E. W. Hodgins, and B. J. Butler. 1991. "Field and Laboratory Evidence of In Situ Biotransformation of Tetrachloroethene to Ethene and Ethane at a Chemical Transfer Facility in North Toronto." In R. E. Hinchee and R. F. Olfenbuttel (Eds.), *On-Site Bioreclamation Processes for Xenobiotic and Hydrocarbon Treatment*, pp. 147-171. Butterworth-Heinemann, Stoneham, MA.

McCarty, P. L., and J. T. Wilson. 1992. "Natural Anaerobic Treatment of a TCE Plume, St. Joseph, Michigan, NPL Site." In U.S. Environmental Protection Agency, *Bioremediation of Hazardous Wastes (abstracts)*, pp. 47-50. EPA/600/R-92/126, Office of Research and Development, Washington, DC.

Salhotra, A. M. , P. Mineart, S. Sharp-Hansen, and T. Allison. 1990. *Multimedia Exposure Assessment Model (MULTIMED) for Evaluating the Land Disposal of Wastes-Model Theory.* U.S. Environmental Protection Agency, Office of Research and Development, Athens, GA.

Vogel, T. M. 1994. "Natural Bioremediation of Chlorinated Solvents." In R. D. Norris, R. E. Hinchee, R. Brown, P. L. McCarty, L. Semprini, J. T. Wilson, D. H. Kampbell, M. Reinhard, E. J. Bouwer, R. C. Borden, T. M. Vogel, J. M. Thomas, and C. H. Ward (Eds.), *Handbook of Bioremediation,.* pp. 201-225. Lewis Publishers, Boca Raton, FL.

Wilson, J. T., J. W. Weaver, and D. H. Kampbell. 1994. "Intrinsic Bioremediation of TCE in Ground Water at an NPL Site in St. Joseph, Michigan." In U.S. Environmental Protection Agency, *Symposium on Intrinsic Bioremediation of Ground Water,* pp. 154-160. EPA/450/R-94/515, Office of Research and Development, Washington, DC.

Intrinsic Biodegradation of Trichloroethene and Trichloroethane in a Sequential Anaerobic-Aerobic Aquifer

Evan Cox, Elizabeth Edwards,
Leo Lehmicke, and David Major

ABSTRACT

Trichloroethene (TCE) and 1,1,1-trichloroethane (1,1,1-TCA) are being intrinsically biodegraded in the groundwater near a former unlined septic-waste lagoon (former lagoon) at a site in Sacramento, California. Although the groundwater at the site is predominantly aerobic, groundwater in the immediate vicinity of the former lagoon is anaerobic. In the anaerobic zone, TCE and 1,1,1-TCA are being sequentially dechlorinated to ethene and ethane. However, the rate of dechlorination is not sufficient to prevent transport of TCE, 1,2-dichloroethene (1,2-DCE), vinyl chloride (VC), 1,1-DCA, and chloroethane (CA) from the anaerobic zone. In the aerobic zone, the lateral distributions of VC and CA are considerably smaller than the lateral distributions of the higher chlorinated volatile organic compounds (VOCs). Because VC and CA are more mobile in groundwater than the higher chlorinated VOCs, and therefore should have greater lateral distributions in the groundwater, we conclude that VC and CA are being biodegraded in the aerobic zone. These results provide field evidence for the intrinsic biodegradation of TCE and 1,1,1-TCA to nonchlorinated end products via two processes: (1) complete reductive dechlorination to ethene and ethane under anaerobic conditions; and (2) reductive dechlorination to VC or CA under anaerobic conditions, followed by biodegradation of VC and CA under aerobic conditions.

INTRODUCTION

Field and laboratory studies have shown that chlorinated VOCs can be biodegraded under both anaerobic and aerobic conditions. For example, Major et al. (1991, 1994), Fiorenza et al. (1994), and Wilson et al. (1994) have shown that chlorinated VOCs can be intrinsically biodegraded to ethene and ethane under anaerobic conditions by indigenous methanogenic, acetogenic, and sulfate-reducing bacteria. Under aerobic conditions, Semprini et al. (1990), Hopkins et al. (1993), and Nelson et al. (1988) have shown that chlorinated VOCs can be cometabolically biodegraded by methanotrophs, phenol-oxidizers, and toluene-oxidizers, whereas Davis and Carpenter (1990) and van den Wijngaard et al. (1992) have shown that certain di- and monochlorinated VOCs can be used directly as carbon and energy sources.

TCE and 1,1,1-TCA and their respective breakdown products (1,2-DCE and VC for TCE; 1,1-DCA and CA for 1,1,1-TCA) have been detected in the groundwater near a former lagoon at a site in Sacramento, California. TCE and 1,1,1-TCA were contaminants in septic waste disposed of in the former lagoon from the early 1960s until 1979. An examination of the VOC distribution data suggested that biodegradation of the chlorinated VOCs was limiting the migration of selected VOCs in the groundwater. Based on the site data and previous research, we conducted this study to determine if, and to what extent, intrinsic biodegradation of VOCs is occurring, to provide an initial evaluation of the types of anaerobic and aerobic microbial processes that are occurring in the groundwater near the former lagoon (e.g., methanogenesis, acetogenesis, sulfate-reduction, aerobic oxidation), and to evaluate whether in situ bioremediation could be considered in the selection and design of remedial alternatives for the site.

Site Description

The former lagoon was situated in a natural depression located between dredged tailing piles, and had a surface area of approximately 550 m^2 and an approximate depth of 6 m. In 1979, the lagoon was filled with dredge tailings from the surrounding area, and graded flat. The dredge tailings are remnants of historic gold-dredging operations, which have disturbed alluvial sediments to depths of up to 30 m below ground surface (bgs). A sludge layer remains at depths of up to 9 m bgs. The average groundwater velocity in the vicinity of the former lagoon is approximately 0.36 m/d. Water-table elevations have decreased by more than 9 m since 1982, and presently range between 15 and 20 m bgs.

Chlorinated VOCs have been reported in groundwater samples collected from monitoring wells located downgradient from the former lagoon since 1980. The highest concentrations of chlorinated VOCs have consistently been reported in monitoring well 33, located 45 m downgradient from the former lagoon (see Figure 1). In 1980, the total chlorinated VOC concentration in a groundwater sample collected from this well exceeded 100,000 μg/L. VOC concentrations have decreased since 1980. In 1994, the total chlorinated VOC concentration in well 33 was 8,300 μg/L.

METHODS

Groundwater samples were collected from 12 monitoring wells screened in the unconfined aquifer near the former lagoon. A minimum of three casing volumes were purged from each monitoring well using a decontaminated submersible pump, to obtain groundwater samples representative of the in situ groundwater chemistry. Dissolved oxygen (DO) was measured using a DO electrode with an accuracy of 0.1 mg/L. Oxidation-reduction potential (ORP) was measured using an ORP electrode with an accuracy of 1 millivolt (mV). Groundwater was continuously pumped through an in-line flowthrough cell containing the DO and ORP electrodes, and readings were recorded when they stabilized.

Groundwater samples for analysis of VOCs, dissolved hydrocarbon gases (ethene, ethane, and methane), specific carbon sources (alcohols, ethers, and acetic acid), biochemical oxygen demand (BOD), and inorganic parameters (nitrate, sulfate, and sulfide) were collected using dedicated Teflon™ bailers. Samples were transferred directly from the bailers into sample containers. Sample vials containing groundwater for analysis of volatile parameters were filled without headspace. Samples were appropriately preserved, placed on ice, and delivered or shipped express to the analytical laboratories for analysis.

VOCs were analyzed by gas chromatography/mass spectrometry (GC/MS) using United States Environmental Protection Agency (U.S. EPA) Method 8240. Nitrate and sulphate were analyzed by ion chromatography using U.S. EPA Method 300.0. Sulfide was analyzed by U.S. EPA Method 376.1. BOD was analyzed by U.S. EPA Method 405.1. Dissolved hydrocarbon gases were analyzed by gas-phase injection/gas chromatography using a Hewlett Packard 5840A GC with a flame ionization detector (FID). Alcohols and ethers were analyzed by aqueous injection/gas chromatography using a Hewlett Packard 5840A GC with a FID detector. Acetic acid was analyzed by ion chromatography using a Waters Ion Exclusion column with a conductivity detector.

RESULTS AND DISCUSSION

Table 1 presents groundwater chemistry data for five monitoring wells screened in the unconfined aquifer and located on the groundwater flowpath. Well 74 is located upgradient of the former lagoon, wells 1727 and 33 are located in the anaerobic zone immediately downgradient from the former lagoon, and wells 3037 and 3054 are located in the aerobic zone downgradient from the former lagoon. The data for these wells illustrate the changes in groundwater chemistry that occur along the groundwater flowpath in the area of the former lagoon. The groundwater chemistry is discussed below.

Redox Conditions

The results of the chemical analyses indicate that a natural sequential anaerobic-aerobic biological system exists in the groundwater beneath and downgradient from the former lagoon. Groundwater immediately downgradient from the former lagoon (wells 1727 and 33) is predominantly anaerobic, characterized by negative ORP, high methane concentrations, nitrate depletion,

TABLE 1. Groundwater chemistry data for selected monitoring wells screened in the unconfined aquifer and located on the groundwater flowpath.

Parameter Well Location:	74 Upgradient from Former Lagoon	1727 Anaerobic Zone	33 Anaerobic Zone	3037 Aerobic Zone	3054 Aerobic Zone
Redox					
Dissolved Oxygen (mg/L)	0.3	0.5	0.7	1.7	9.0
Oxidation-Reduction Potential (mV)	85	−61	−125	79	289
Volatiles (μg/L)					
Trichloroethene	<5.0[b]	6.1	1400	55	7.5
1,2-Dichloroethene (*cis* and/or *trans*)	<5.0	270	2800	390	69
Vinyl Chloride	<10	1400	2600	<10	<10
Ethene	<1.0	180	112	<1.0	<1.0
1,1,1-Trichloroethane	<5.0	<5.0	<5.0	<5.0	<5.0
1,1-Dichloroethane	<10	400	1400	140	5.9
Chloroethane	<10	87	<50	<10	<10
Ethane	<1.0	9.2	<1.0	<1.0	<1.0
Methane	26.6	256	200	38.7	<1.0
Organic and Inorganic Nutrients (mg/L)					
Oxygenates[a]	<1.0	<1.0	<1.0	<1.0	<1.0
Acetic Acid	<5.0	<5.0	<5.0	<5.0	<5.0
Nitrate	0.97	0.36	0.53	8.5	7.5
Sulfate	14	1.4	12	10	5.9
Sulfide	<0.3	<0.3	<0.3	<0.3	<0.3
Biological Oxygen Demand	<1.0	4.0	1.6	<1.0	<1.0

(a) Oxygenates consist of acetone, methanol, tert-butyl alcohol, and methyl-tert-butyl ether.

(b) Analyte was not detected at a concentration greater than the associated quantitation limit.

and DO depletion (Table 1). There appears to be a transition from anaerobic conditions to aerobic conditions between wells 33 and 3037. Groundwater in the aerobic zone (wells 3037 and 3054) is characterized by positive ORP, low methane concentrations, and the presence of DO and nitrate at higher concentrations (Table 1).

Distribution of Selected Chlorinated VOCs, Ethene, and Ethane

Figures 1a to 1d show the lateral distribution of TCE, 1,2-DCE, VC, and ethene in the vicinity of the former lagoon. Figures 2a to 2c show the lateral distribution of 1,1-DCA, CA, and ethane in the vicinity of the former lagoon. 1,1,1-TCA has not been detected in the groundwater above its reporting limit (5 μg/L) since 1986.

The presence of 1,2-DCE, VC, 1,1-DCA, and CA in the groundwater provides evidence that TCE and 1,1,1-TCA are being biodegraded. These degradation products were not used or produced at the site, and therefore their presence can only be attributed to the dechlorination of TCE and 1,1,1-TCA. Ethene and ethane were detected in the groundwater samples collected from monitoring wells located within the anaerobic zone, providing evidence that the chlorinated ethenes and ethanes are being biodegraded to innocuous nonchlorinated end products. However, the rate of dechlorination of the

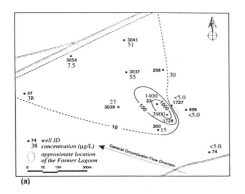

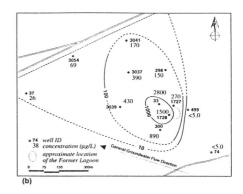

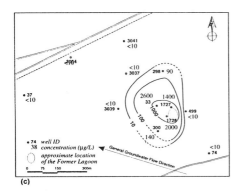

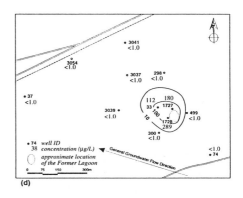

FIGURE 1. Isoconcentration contours of (a) trichloroethene, (b) 1,2-dichloroethene, (c) vinyl chloride, and (d) ethene in the groundwater near the former lagoon (contours dashed where inferred).

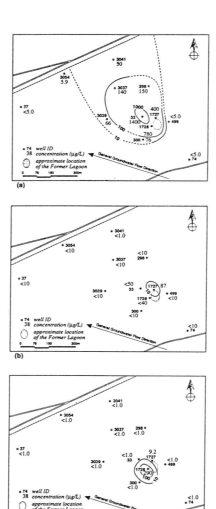

FIGURE 2. Isoconcentration contours of (a) 1,1-dichloroethane, (b) chloroethane, and (c) ethane in the groundwater near the former lagoon (contours dashed where inferred).

chlorinated VOCs to ethene and ethane in the anaerobic zone does not appear
to be fast enough to prevent the transport of some the chlorinated VOCs from
the anaerobic zone to the aerobic zone.

In the aerobic zone, the lateral distributions of VC and CA are considerably
smaller than the lateral distributions of the higher chlorinated VOCs. VC and
CA are more mobile in groundwater than the higher chlorinated VOCs, and
therefore should have greater lateral distributions in the groundwater. Because

the lateral distribution of VC and CA in the groundwater is small relative to 1,2-DCE, TCE, and 1,1-DCA, we propose that VC and CA are being biodegraded in the aerobic zone.

The historical VOC data for monitoring well 33, located in the anaerobic zone approximately 45 m downgradient from the former lagoon, and well 3037, located in the aerobic zone approximately 275 m downgradient from the former lagoon, provide support for our hypothesis that VC and CA are being biodegraded in the aerobic zone. VC, 1,2-DCE, TCE, CA, and 1,1-DCA have all been detected in groundwater samples collected from well 33 since 1980. In contrast, only TCE, 1,2-DCE, and 1,1-DCA have been detected in groundwater samples from well 3037, since its installation in 1986. VC and CA have not been detected at concentrations above their detection limit of 10 μg/L in well 3037. Based on the average linear groundwater velocity near the former lagoon of 0.36 m/d, the average groundwater travel time between monitoring wells 33 and 3037 would be approximately 640 days. Factoring in retardation (calculations presented in Table 2), the travel times for the VOCs from well 33 to well 3037 would range from 649 days for VC to 1,070 days for TCE (Table 2). Based on the range of travel times and the concentrations of the VOCs in well 33, each of the VOCs should be present at concentrations well above the detection limit (10 μg/L) in the groundwater at well 3037. Because VC and CA are not present in the groundwater at well 3037, and because they are known to be rapidly biodegraded under aerobic conditions, we conclude that VC and CA are being biodegraded in the aerobic zone.

TABLE 2. Calculated retardation factors and travel times from well 33 to well 3037 for VC, CA, 1,1-DCA, 1,2-DCE, and TCE.

Parameter	K_{oc}[a]	Kd[b]	Retardation[c]	Travel Time from well 33 to well 3037 (days)[d]
Vinyl Chloride	2.5	0.0025	1.01	649
Chloroethane	3.2	0.0032	1.02	651
1,1-Dichloroethane	30	0.030	1.16	742
1,2-Dichloroethene	59	0.059	1.31	841
Trichloroethene	126	0.126	1.67	1070

(a) K_{oc} calculated from Log K_{oc} values in Montgomery and Welkom (1990).
(b) $Kd = K_{oc} \cdot f_{oc}$ (assuming an f_{oc} value for the unconfined aquifer of 0.1%).
(c) Retardation = 1+(bulk density/porosity)• Kd (assuming a bulk density of 1.6 g/cm^3, and a porosity of 0.30).
(d) Based on the average groundwater travel time of 640 days from well 33 to well 3037 (using the calculated average groundwater velocity of 0.36 m/d, and the distance between the wells of 230 m).

Distribution of Organic and Inorganic Nutrients

The reductive dechlorination of the chlorinated VOCs in the anaerobic zone has probably been promoted by organic substrates from the septic waste. However, the specific carbon substrates that are being used by the anaerobic microorganisms have not been identified. Oxygenates such as acetone and methanol, common groundwater contaminants that can serve as growth substrates and electron donors in dechlorination (Major et al. 1991, 1994), were not detected in any groundwater samples. Acetic acid, a product of the metabolism of simple carbon substrates by acetogenic bacteria, was not detected in any of the groundwater samples. Methanogenesis is occurring in the anaerobic zone, as indicated by the high concentrations (up to 1,570 μg/L) of methane in the groundwater samples from the anaerobic zone. Sulfate reduction may be occurring in the anaerobic zone, as indicated by the low concentration (1.4 mg/L) of sulfate in well 1727; however, sulfide was not detected. The BOD in groundwater samples collected from monitoring wells located in the anaerobic zone was lower than expected (maximum of 4.5 mg/L), given the proximity of the wells to the former lagoon. The low BOD suggests that microbial activity and dechlorination may be limited by the availability of suitable electron donors.

CONCLUSIONS

The intrinsic biodegradation of chlorinated VOCs in groundwater has generated significant interest from researchers and regulatory agencies. This study provides field evidence for the intrinsic biodegradation of TCE and 1,1,1-TCA to nonchlorinated end products via two processes: (1) complete reductive dechlorination to ethene and ethane under anaerobic conditions; and (2) reductive dechlorination to VC or CA under anaerobic conditions, followed by biodegradation of VC and CA under aerobic conditions. In the system studied, the rate of reductive dechlorination is not sufficient to completely control the migration of selected VOCs. Future laboratory and field studies will be conducted to determine if biodegradation can be enhanced with a minimum of intervention, and whether in situ bioremediation is an appropriate remedial alternative for the site.

REFERENCES

Davis, J. W., and C. L. Carpenter. 1990. "Aerobic Biodegradation of Vinyl Chloride in Groundwater Samples." *Appl. Environ. Microbiol.* 56:3878-3880.

Fiorenza, S., E. L. Hockman Jr., S. Szojka, R. M. Woeller, and J. W. Wigger. 1994. "Natural Anaerobic Degradation of Chlorinated Solvents at a Canadian Manufacturing Plant." In R.E. Hinchee, A. Leeson, L. Semprini, and S.K. Ong (Eds.), *Bioremediation of Chlorinated and Polycyclic Aromatic Hydrocarbon Compounds*. Lewis Publishers, Boca Raton, FL, pp. 227-286.

Hopkins, G. D., L. Semprini, and P. L. McCarty. 1993. "Microcosm and In Situ Field Studies of Enhanced Biotransformation of Trichloroethylene by Phenol-Utilizing Microorganisms." *Appl. Environ. Microbiol. 59*:2277-2285.

Major, D. W., E. E. Cox, E. Edwards, and P. W. Hare. 1994. "The Complete Dechlorination of Trichloroethene to Ethene under Natural Conditions in a Shallow Bedrock Aquifer Located in New York State." *Symposium on Intrinsic Bioremediation of Groundwater.* United States Environmental Protection Agency, EPA/540/R-94/515, August 1994, Denver, CO.

Major, D. W., E. H. Hodgins, and B. H. Butler. 1991. "Field and Laboratory Evidence of In Situ Biotransformation of Tetrachloroethene to Ethene and Ethane at a Chemical Transfer Facility in North Toronto." In R. E. Hinchee and R. F. Olfenbuttel (Eds.), *On-Site Bioreclamation.* Butterworth-Heinemann, Stoneham, MA, pp. 147-171.

Montgomery, J. H., and L. M. Welkom. 1990. *Groundwater Desk Reference.* Lewis Publishers, Chelsea, MI.

Nelson, M. J. K., S. O. Montgomery, and P. H. Pritchard. 1988. "Trichloroethylene Metabolism by Microorganisms that Degrade Aromatic Compounds." *Appl. Environ. Microbiol. 54*:604-606.

Semprini, L., P. V. Roberts, G. D. Hopkins, and P. L. McCarty. 1990. "A Field Evaluation of In Situ Biodegradation of Chlorinated Ethenes, Part 2. Results of Biostimulation and Biotransformation Experiments." *Groundwater 28*:715-727.

van den Wijngaard, A. J. K. W. van der Kamp, J. van der Ploeg, F. Pries, B. Kazemier, and D. B. Janssen. 1992. Degradation of 1,2-Dichloroethane by *Ancylobacter aquaticus* and Other Facultative Methylotrophs." *Appl. Environ. Microbiol. 58*: 976-983.

Wilson, J. T., J. W. Weaver, and D. H. Kampbell. 1994. "Intrinsic Bioremediation of TCE in Ground Water at an NPL Site in St. Joseph, Michigan." *Symposium on Intrinsic Bioremediation of Groundwater.* United States Environmental Protection Agency, EPA/540/R-94/515, August 1994, Denver, CO.

Inferring Biodegradation Processes for Trichloroethene from Geochemical Data

*Peter R. Guest, Leigh A. Benson,
and Tony J. Rainsberger*

ABSTRACT

An increasing number of laboratory investigations and field case studies show that natural biodegradation processes have a significant effect on contaminant fate and transport in groundwater. The importance of these processes should be quantitatively characterized so that practical and cost-effective remedial strategies can be developed. A protocol document is being developed by the U.S. Air Force documenting the effects of natural biodegradation processes on the fate of dissolved fuel-related contaminants. The same protocol developed for fuel-related contaminants was applied at three different sites contaminated with chlorinated aliphatic hydrocarbons (CAH) at a facility in Colorado as a first-line indication of CAH degradation. Data on the chemical characteristics of the CAH compounds and the affected environmental medium can be used to infer the type of microbial degradation processes that may be influencing contaminant fate at the three sites. These data suggest that dissolved CAHs may be biodegrading via anaerobic cometabolic processes at two of the sites, but not biodegrading at the third site. This information provides insight into the natural biodegradation processes that appear to be occurring at each site. Understanding these processes is vital for evaluating remedial alternatives that may involve enhancing the natural biodegradation of CAHs.

INTRODUCTION

CAHs are used in a multitude of industrial and agricultural applications. For example, trichloroethene (TCE) has been and is still commonly used as a solvent and cleaning agent. Improper waste disposal practices and inadvertent chemical spills have contaminated groundwater at many sites. These dissolved chemicals may impact groundwater and drive remediation strategies for a site.

Laboratory studies suggest that CAH compounds such as TCE can be biodegraded under both aerobic and anaerobic conditions, provided the appropriate microbial consortia and geochemical conditions are present (Wilson and Wilson 1985; Nelson et al. 1986; Freedman and Gosset 1989; DiStefano et al. 1991; McCarty and Semprini 1994). These studies provide a framework for defining and understanding the potential biochemical degradation mechanisms that may be operating in soil and groundwater at a contaminated site. However, it is difficult (if not impossible) to simulate site-specific field conditions in laboratory studies. Therefore, it is necessary to collect and evaluate site-specific analytical data to determine if contaminant biodegradation processes are occurring. This site-specific information can then be coupled with laboratory data to understand if and how CAH compounds are biodegrading at a site. Documenting the occurrence of natural CAH biodegradation processes plays a significant role in defining the types of remedial technologies that may be appropriate for a site.

CAH BIODEGRADATION PROCESSES

The mechanism of biochemical transformation of CAH compounds depends on the chemical nature of the CAH compound and the environmental conditions. CAH biodegradation can occur via microbially mediated reduction-oxidation (redox) reactions where the CAH compound is further oxidized and another chemical species is further reduced. Degradation of CAH compounds is a cometabolic reaction in which degradation of a primary substrate is required to supply the necessary energy and carbon requirements. CAH compounds that are less chlorinated, such as vinyl chloride (VC), are more likely to undergo oxidation reactions. Oxidation of the less chlorinated CAH compounds is most favorable under aerobic conditions. In contrast, the CAH compounds can also be reductively dehalogenated by using electrons donated by another chemical species. More chlorinated CAH compounds such as TCE and tetrachloroethene (PCE) are more amenable to progressive dechlorination reactions. These reactions are most favorable under anaerobic and reducing conditions. Data collected from two of the three investigated sites indicate that anaerobic and reducing conditions are present in contaminated groundwater.

CAH compounds can be progressively dechlorinated under anaerobic conditions. Reductive dechlorination involves the removal of a chlorine atom from the CAH while an appropriate electron donor is degraded. TCE can be reductively dechlorinated to 1,2-*cis*-dichloroethene (*cis*-DCE), 1,2-*trans*-DCE (*trans*-DCE), and VC, and, in certain fortuitous circumstances, to ethene and/or ethane. However, dechlorination of VC to ethene is typically the rate-limiting step for complete degradation of TCE to ethane. Of the DCE isomers, laboratory results suggest that *cis*-DCE dominates over *trans*-DCE (Parsons et al. 1984).

A broad variety of microbial consortia apparently possess the enzymatic capabilities to reductively dechlorinate CAH compounds. The effectiveness of this metabolic process is usually limited by the availability of appropriate electron donors (primary substrate), such as low-molecular-weight organic compounds, gaseous hydrocarbons, toluene, or other volatile fatty acids. The anaerobic processes are not as effective as aerobic oxidation in terms of complete transformation, and CAH compounds are usually only partially dechlorinated under anaerobic reducing conditions. In most situations, other processes must be operating, in addition to anaerobic reductive dechlorination, to promote total dechlorination of highly chlorinated ethylenes into ethene and/or ethane (McCarty and Semprini 1994).

CONTAMINANT PATTERNS

The groundwater beneath three sites at an Air Force base in Colorado is contaminated with the CAH compounds TCE, *cis*-DCE, *trans*-DCE, and VC. Two of the sites exhibit residuals of fuel contamination in site media. Each site had different historical uses. Groundwater sampling and analysis were recently conducted at these sites as part of a remedial investigation/feasibility study to characterize the nature and extent of contamination and to investigate the potential for natural CAH biodegradation processes. A brief synopsis of site history, governing hydrogeologic conditions, and relevant contaminant patterns for each site follows.

Fire Training Zone

The groundwater at the Fire Training Zone was most likely contaminated with CAH compounds during fire training exercises conducted between 1946 and 1965. Waste fuels and solvents were typically poured onto airplane fuselages or into a bermed area, and ignited. Minimal quantities of the jet fuel JP-4 was also used in these fire training exercises. The groundwater underneath the Fire Training Zone is unconfined and occurs in the siltstone and claystone that comprises the bedrock formation. The depth to groundwater at this site ranges from 12 to 17 ft (3.7 to 5.2 m) below ground surface (bgs).

Recent sampling and analytical results showed elevated concentrations of the CAH compound TCE in the groundwater (Table 1). No evidence of fuel residuals (i.e., benzene, toluene, ethylbenzene, or xylenes [BTEX] or total petroleum hydrocarbons [TPH]) was detected in recent groundwater samples collected at the Fire Training Zone. No CAH compounds or TPH were found in soil samples taken from the site. Limited BTEX (less than 10 mg/kg) contamination was detected in site soils.

Auto Hobby Shop

This site was used to support automobile maintenance, repair, parts cleaning, and washing. Several sources of contamination have been identified,

TABLE 1. Fire Training Zone: Chlorinated aliphatic hydrocarbon and electron acceptor concentrations.

Wells	TCE (µg/L)	c-DC (µg/L)	t-DCE (µg/L)	VC (µg/L)	O_2 (mg/L)	NO_3 (mg/L)	SO_4 (mg/L)
FT-1	3 U[a]	1U	2 UJ[b]	1 UJ	3.2	7.7	104.5
FT-11	5	1 U	2 U	1 U	4.3		165.3
FT-12	2	1 U	2 U	1 U	5.5	11	161.6
FT-13	16	1 U	2 U	1 U	5.3	24	110.1
FT-17	85	1 U	2 U	1 U	3.8	14	166.6

(a) U is undetected.
(b) UJ is undetected, estimated quantity.

including a former 1,000-gal (3,785-L) aboveground storage tank (AST) used to store spent solvent from cleaning operations, an oil/water separator, a former drum storage shed, and a 550-gal (2,082-L) waste-oil AST. The groundwater beneath the site is unconfined and occurs in unconsolidated sand and silt deposits. Depth to groundwater ranges from 6.5 to 9.5 ft (2 to 2.9 m) bgs.

Relatively low concentrations (48 µg/L) of *cis*-DCE, *trans*-DCE, and VC were recently detected in groundwater at the site (Table 2). Low concentrations of toluene, ethylbenzene, and xylenes (less than 5 mg/kg) were detected in site groundwaters and soils. Oil and grease were detected in site soils at a maximum concentration of 26,000 mg/kg, but not in site groundwaters. The presence of CAH and petroleum hydrocarbons in both soil and groundwater is consistent with historical use records for the site.

Fly Ash Disposal Area

This site was reportedly used for disposal of fly ash and uncombusted coal waste generated within combustion-furnace heating facilities from 1940 to 1948. Early records also suggest that industrial wastes such as greases and solvents associated with aircraft maintenance were disposed of into maintenance-shop drains, which ultimately discharge within the southern portion of the Fly Ash Disposal Area. The outfall of these drains is immediately upgradient of the highest detected concentrations of CAHs, and has been determined to be the suspected contamination source area. Elevated concentrations of TCE, *cis*-DCE, *trans*-DCE, and VC were found immediately downgradient of the storm drain outfall (Table 3) in samples collected using cone penetrometer (piezo-cone/Hydrocone) techniques (CPT), or through the installation of permanent monitoring wells using conventional augering techniques.

The groundwater beneath the site is unconfined and occurs in the unconsolidated sands and silts near the suspected source area, and within the silt-stone and claystone of the bedrock formation downgradient of the site. The depth to groundwater at this site ranges from 5 to 25 ft (1.5 to 7.6 m) bgs.

TABLE 2. Auto Hobby Shop: Chlorinated aliphatic hydrocarbon and electron acceptor concentrations.

Wells	TCE (μg/L)	c-DC (μg/L)	t-DCE (μg/L)	VC (μg/L)	O_2 (mg/L)	NO_3 (mg/L)	SO_4 (mg/L)
AHS-1	3 U[a]	1U	2 U	1 U	4.6	0.6	164.8
AHS-2	0 U	1 U	2 U	1 U	2.7	0.5	165.2
AHS-3	1 U	1 U	2 U	1 U	6.1	8.6	164.9
AHS-4	1 U	48	23	2.4	5.3	0.1	135.4
S-5-1	3 U	1 J[b]	2 UJ[c]	1 UJ	3.4	0.7	164.2

(a) U is undetected.
(b) J is estimated quantity.
(c) UJ is undetected, estimated quantity.

TABLE 3. Fly Ash Disposal Area: Chlorinated aliphatic hydrocarbon concentrations.

Sample ID	Trichloro-ethene	Cis-1,2-DCE	Trans-1,2-DCE	Vinyl Chloride
		Source Area		
CP84-10'	3,282.00	177.01	25.84	ND[a]
CP84-15'	93,160.00	76.55	23.37	ND
CP100-12'	107,925.00	969.55	22.77	39.21
CP86-12'	24,815.00	15,100.00	41.20	314.50
CP82-14.5'	37,233.00	11,929.50	ND	978.00
CP77-17'	34,072.00	8,862.00	27.45	730.00
CP94-15'	10,640.00	2,299.90	8.13	286.42
CP95-12'	47.19	56.85	14.88	ND
		Downgradient		
CP16-35'	253.00	19.00	1J	ND
CP17-29'	413.00	28.00	2J	ND
CP18-35'	540.00	70.00	7.00	6.00
CP19-35'	14.00	ND	ND	18.00
FA-3	290.00	20J	1.00	ND
		Leading Edge of Plume		
CP22-35'	ND	ND	ND	14.00
CP23-17'	ND	ND	ND	16.00
CP33-33'	ND	ND	ND	12.00
CP35-23'	ND	ND	ND	17.00
CP38-49'	ND	ND	ND	10.00
CP47-42'	70.00	3J	ND	3J

(a) ND = target analyte not detected.

CHEMICAL INDICATIONS OF
CAH BIODEGRADATION PROCESSES

The Air Force Center for Environmental Excellence (AFCEE) is currently sponsoring an effort to develop a technical protocol document on geotechnical sampling techniques, analytical methods, and analysis approaches relevant to documenting the natural biodegradation of dissolved petroleum hydrocarbons (Wiedemeier et al. 1994). This protocol was implemented with certain chemical-specific modifications at each of the three sites to test whether geochemical data could be used to infer biochemical processes of CAH degradation.

Fire Training Zone

Table 1 summarizes the CAH concentrations, and electron acceptor concentrations measured within groundwater wells at the Fire Training Zone. PCE and TCE were detected in samples collected from site groundwaters, but *cis*-DCE, *trans*-DCE, and VC were not detected in these groundwater samples. Concentrations of electron acceptors measured in groundwater samples collected from contaminated wells do not appear to be reduced relative to concentrations of electron acceptors measured in samples collected from uncontaminated wells. The absence of the DCE isomers and VC indicate that reductive dechlorination of TCE is not currently occurring. Additionally, site conditions appear to be oxidizing, since oxygen, nitrate, and sulfate concentrations do not appear to be reduced within contaminated site groundwater relative to uncontaminated site groundwaters. Anaerobic biological reductive dechlorination is not known to occur under highly oxidized conditions. The apparent oxidizing conditions and lack of the DCE isomers and VC in site groundwaters suggest that anaerobic reductive dechlorination is not occurring at this site. Thus, although PCE could be reductively dechlorinated to TCE, redox conditions are not favorable for the transformation at this site. More likely, PCE and TCE were both introduced into this site, which is consistent with the historical use of the site.

Auto Hobby Shop

Table 2 summarizes CAH analytical results and electron acceptor field measurements in groundwater samples collected at the Auto Hobby Shop. TCE was not detected in laboratory analytical groundwater samples collected at the Auto Hobby Shop. In the laboratory analytical groundwater samples collected from well AHS-4, *cis*-DCE, *trans*-DCE, and VC were detected. Historical evidence suggests that DCE was not used in normal site operations. Therefore, the presence of *cis*-DCE, *trans*-DCE, and VC implies that anaerobic reductive dechlorination of TCE may have occurred in the past. Cometabolism of TCE may have occurred in the presence of petroleum hydrocarbons which would serve as the electron donors (primary substrate). The relative concentrations of *cis*-DCE and *trans*-DCE appear to be consistent with concentrations observed in laboratory studies of reductive dechlorination of TCE. Dissolved oxygen (DO) concentrations within the contaminated well, AHS-4, do not appear to be reduced relative to other DO concentrations measured within uncontaminated wells at the site. However, nitrate and sulfate concentrations measured in groundwater collected from well AHS-4 appear to be reduced relative to the nitrate and sulfate concentrations measured in groundwater samples collected from uncontaminated wells in the same area. Available site data demonstrate three trends commonly associated with reductive dechlorination: the presence of less chlorinated ethenes, which are found when reductive dechlorination of TCE occurs; degradation breakdown products at relative concentrations commonly found when reductive dechlorination of TCE occurs; and reduced concentrations of nitrate and sulfate. Although elevated concentrations of DO in contaminated groundwater typically indicate oxidizing conditions, which would preclude the occurrence of reductive dechlorination, the presence of reduced electron acceptors suggests more reducing conditions prevail. More complete groundwater data are necessary to determine if reductive dechlorination of CAHs is occurring at the Auto Hobby Shop.

Fly Ash Disposal Area

Site characterization data for this site are more extensive than for either the Fire Training Zone or the Auto Hobby Shop. Concentrations of TCE detected by groundwater sampling conducted at the Fly Ash Disposal Area are presented in Figure 1. Significant concentrations of TCE and associated degradation breakdown products detected in groundwater samples collected from the Fly Ash Disposal Area suggest that reductive dechlorination is occurring (Table 3). Data from a limited number of groundwater samples indicate that electron acceptors within the CAH plume are reduced relative to surrounding uncontaminated groundwater (Table 4). Measured concentrations of the oxygen, nitrate, and (to a lesser degree) sulfates appear to be lower within CAH-contaminated groundwater samples versus uncontaminated groundwater samples. Of the DCE isomers detected within site groundwater, *cis*-DCE concentrations are significantly higher than *trans*-DCE. This observation is consistent with laboratory studies, which indicate that the *cis*-DCE isomer typically occurs in greater concentrations than the *trans*-DCE isomer when TCE undergoes reductive dechlorination. DCE concentrations measured in contaminated groundwater samples are significantly greater than VC concentrations. Again, this observation is consistent with theoretical decreases in reductive dechlorination rates observed with CAHs with lower degrees of chlorination. Additionally, when anaerobic reductive dechlorination of chlorinated ethenes occurs, associated breakdown products migrate at different rates, and these breakdown products begin to separate into regions within the plume. Migration rates predicted using octanol/water coefficients indicate that VC should lead the plume, followed by the DCE isomers, and then TCE (Montgomery and Welkom 1989). Observed spatial relationships between TCE and associated breakdown products follow this expected spatial relationship, which should develop when

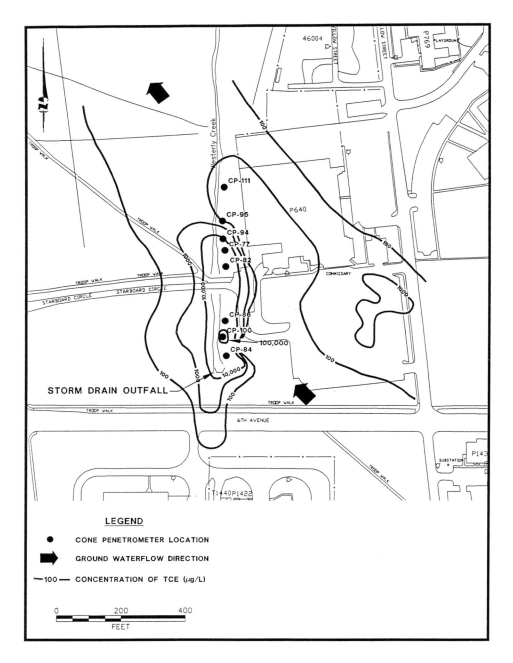

FIGURE 1. Concentration of TCE in groundwater source area at the Fly Ash Disposal Area.

TABLE 4. Fly Ash Disposal Area: Electron acceptor concentrations.

Sample ID	Location	Oxygen (mg/L)	Nitrate (mg/L)	Sulfate (mg/L)
FA-2	background	7.8	5.3	170.6
D-6-2	source	5.6	0.7	156.9
FA-3	downgradient	5.2	0.8	147.1
FA-4	downgradient	2.6	0.2	165.4

TCE is undergoing reductive dechlorination (Figure 2). Observed spatial relationships between plume CAHs, the relative concentrations of breakdown products present, and the apparent depletion of electron acceptors within the plume indicate that reductive dechlorination is occurring at the Fly Ash Disposal Area.

IMPLICATIONS

Fire Training Zone

Anaerobic CAH degradation does not appear to be occurring at the Fire Training Zone. Therefore, under the current site conditions, anaerobic degradation is not recommended to be included in a proposed remedial action plan. However, evaluations of aerobic and abiotic degradation pathways should be completed. Site contaminant concentrations of TCE and PCE are relatively low and do not appear to pose a threat to downgradient receptors. Therefore, they could be remediated solely by naturally occurring physical, chemical, and biological processes. A thorough evaluation of these processes in groundwater at the Fire Training Zone would provide valuable information to support implementation of the most cost-effective remedial alternative.

Auto Hobby Shop

Currently available site data are not sufficient to definitively determine if anaerobic reductive dechlorination is occurring in groundwater beneath the Auto Hobby Shop. Additional groundwater sampling is required to determine if anaerobic degradation of the low concentrations of CAHs present beneath the Auto Hobby Shop is occurring. Further evaluation of the naturally occurring physical, chemical, and biological processes could determine if these processes are sufficient alone to minimize potential risks to human health and the environment at this site.

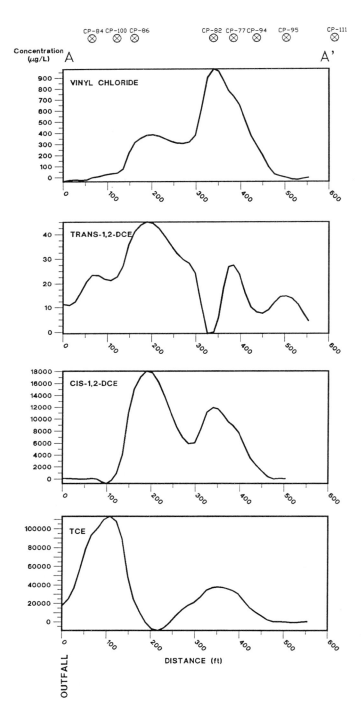

FIGURE 2. Spatial distrubution of TCE degradation products at the Fly Ash Disposal Area.

Fly Ash Disposal Area

Anaerobic CAH degradation appears to be occurring at the Fly Ash Disposal Area. It is necessary to establish degradation rates before reductive dechlorination can be effectively proposed as part of a remedial action plan. The relatively high concentrations of CAHs detected in site groundwater could prevent naturally occurring physical, chemical, and biological processes from being the only mechanism used to reduce potential risks to human health and the environment at this site. However, the reductive dechlorination within the CAH plume could be capitalized on within proposed remedial action plans. For example, the leading edge of the CAH plume consists primarily of VC. Therefore, it would be possible to exploit the mobility of VC and its amenability to aerobic degradation within proposed remedial actions. To target the most mobile chemical, low-cost remedial alternatives could include air sparging, methane introduction, and soil vapor extraction at the leading edge of the plume to simultaneously remove and aerobically degrade VC. This type of scenario would not be feasible without decreasing the degree of chlorination of CAHs in site groundwater through naturally occurring reductive dechlorination.

REFERENCES

DiStefano, T.D., J.M. Gossett, and S.H. Zinder. 1991. "Reductive dechlorination of high concentrations of tetrachloroethene to ethene by an anaerobic enrichment culture in the absence of methanogenesis." *Applied and Environmental Microbiology* 57(8):2287-2292.

Freedman, D.L., and J.M. Gossett. 1989. "Biological reductive dechlorination of tetrachloroethylene and trichloroethylene to ethylene under methanogenic conditions." *Applied and Environmental Microbiology* 55(4):1009-1014.

McCarty, P.L., and L. Semprini. 1994. "Ground-water treatment for chlorinated solvents." In R.D. Norris et al. (Eds.), *Handbook of Bioremediation*, Ch. 5. Lewis Publishers, Boca Raton, FL.

Montgomery, J.H., and Welkom, L.M. 1989. *Groundwater Chemicals Desk Reference.* Lewis Publishers, Inc., Chelsea, MI.

Nelson, M.J.K., S.O. Montgomery, E.J. O'Neille, and P.H. Pritchard. 1986. "Aerobic metabolism of trichloroethylene by a bacterial isolate." *Applied and Environmental Microbiology* 52(2):949-954.

Parsons, F., P.R. Wood, and J. DeMarco. 1984. "Transformation of tetrachloroethylene and trichloroethene in microcosms and groundwater." *Journal of American Water Works Association* 72(2):56-59.

Wiedemeier, T.H., D.C. Downey, J.T. Wilson, D.H. Kampbell, R.N. Miller, and J.E. Hansen. 1994. "Draft technical protocol for implementing the intrinsic remediation with long-term monitoring option for natural attenuation of dissolved-phase fuel contamination in groundwater." Air Force Center for Environmental Excellence, Brooks Air Force Base, TX.

Wilson, J.T., and B.H. Wilson. 1985. "Biotransformation of trichloroethylene in soil." *Applied and Environmental Microbiology* 49(1):242-243.

Intrinsic and Accelerated Anaerobic Biodegradation of Perchloroethylene in Groundwater

Ronald J. Buchanan, Jr., David E. Ellis,
J. Martin Odom, Paul F. Mazierski, and Michael D. Lee

ABSTRACT

The DuPont Niagara Falls Plant is located in a heavily industrialized area of Niagara Falls, New York, adjacent to the Niagara River. The plant has been in continuous operation since 1898 and manufactured various organic and inorganic chemicals. Chlorinated solvents were produced from 1930 to 1975 at the plant. Numerous hydrogeologic investigations have described the subsurface hydrogeology and indicated that the groundwater underlying the plant was impacted by a variety of chlorinated aliphatic hydrocarbons in a wide range of concentrations. DuPont initiated in-field evaluations to determine whether biological reductive anaerobic dechlorination was occurring naturally and, if so, whether such dechlorination could be enhanced in situ. A field program was subsequently implemented in a preselected area of the plant through use of an in situ "borehole bioreactor" to attempt to stimulate indigenous biological reductive dechlorination of chlorinated aliphatics by the addition of yeast extract (substrate) and sulfate (electron acceptor). At this location, a very active microbial population developed, which reduced the in situ concentrations of chlorinated aliphatic compounds by more than 94%, but did not increase the typical biological degradation products. This may have been due to an alternative biological degradation pathway or to very rapid biological kinetics. Efforts to elucidate this mechanism have been initiated under a separate laboratory program.

INTRODUCTION

The Niagara Falls plant is underlain by relatively thin (4.6 to 6.1 m), unconsolidated overburden deposits (primarily fill material termed the "A zone") and deeper fractured dolomitic bedrock. Groundwater is encountered in the overburden and underlying bedrock units. Hydraulic conductivities (10E-04 cm/s) and groundwater flow velocities (0.03 m/day) are generally low in these units.

Groundwater flow in the A zone overburden and A zone bedrock is south toward the Niagara River. Flow in the deeper bedrock zones depends on the spacing, orientation, and interconnection of individual fractures, but generally is toward the north and northeast.

Figure 1 shows the layout of the plant and location of groundwater monitoring wells. Previous groundwater monitoring data indicated the presence of various chlorinated aliphatic compounds. Closer inspection of these data revealed not only the distribution of contaminants, but the potential presence of indigenous microbial reductive dechlorination in specific areas of the plant. Analysis of isopleths indicated areas of *cis*-dichloroethene (DCE) enrichment relative to

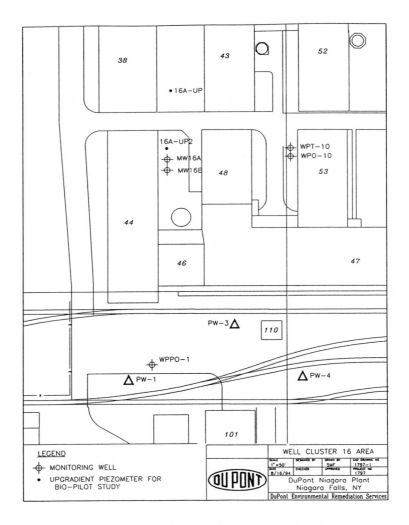

FIGURE 1. Niagara Falls plant layout showing MW-16A.

trans-1,2 DCE and 1,1-DCE above and beyond what would be predicted from manufactured PCE or TCE, or that might randomly occur. Ethene also was detected at elevated concentrations. Based on these results and the work of Beeman et al. (1994) and Beeman (1994) (after evaluating similar plots of other daughter products) DuPont decided to conduct additional in-field tests to confirm this preliminary evidence.

EXPERIMENTAL PROCEDURES AND MATERIALS

Field Studies

The A zone was selected (groundwater monitoring well MW-16A) for the in-field testing based on the concentration range of chlorinated aliphatics (10 mg/L to 20 mg/L), total volatile organic compounds (VOCs), and the shallow depth of the unit. Initially, yeast extract and sulfate were added directly to MW-16A on alternate days each week and carefully mixed downhole with a surge block.

Surging (mixing) consisted of adding a predissolved substrate slurry to the well and gently agitating the water column with a surge block. The surge block itself consisted of three stainless steel, movable flanges (7.62 cm diameter) mounted via set screws to a center galvanized pipe or "handle" (1.27 cm diameter) that contained threaded ends. The flanges were positioned approximately 0.33 m apart from each other. The flanges contained a center aperture to accommodate the handle and four apertures throughout their diameter to reduce fluid drag and lessen agitation aggressiveness. The handle could be lengthened or shortened by attaching various lengths of pipe to the threaded ends.

Addition of amendments occurred during the early morning hours and sampling occurred during the late afternoon to allow the well time to equilibrate.

The groundwater from the well was sampled weekly and analyzed for VOCs, total organic carbon (TOC), chemical oxygen demand (COD), pH, redox potential (Eh), temperature, dissolved oxygen (DO), sulfate, nitrogen, and phosphate. Field parameters (pH, Eh, DO, temperature) were measured downhole using a Hydrolab Model H2OG probe and analyzer.

RESULTS

Analysis of groundwater from MW-16A indicated the following levels of chlorinated aliphatic compounds at the beginning of this study: dichloromethane (DCM), 0.059 mg/L; 1,1-dichloroethane (1,1-DCA) 0.061 mg/L; *cis*-1,2-dichloroethene (*c*-DCE), 0.064 mg/L; 1,1,1-trichloroethane (1,1,1-TCA), 0.088 mg/L; trichloroethene (TCE), 7.4 mg/L; and tetrachloroethene (PCE), 0.11 mg/L.

Additional analytes indicated that the groundwater was anaerobic with a dissolved oxygen content of less than 1.0 mg/L and a redox potential, Eh, of −165 mv. The initial pH was 11.4. The groundwater also contained 370 mg/L sulfate, 0.05 mg/L nitrate, and 0.19 mg/L total phosphate. The initial TOC and initial COD of the groundwater were, 59 mg/L (TOC) and 220 mg/L (COD), respectively.

MW-16A was initially dosed with 650 g of yeast extract on Monday (days 1, 8, and 15) and 1,300 g of sulfate on Tuesday. After day 19, the yeast extract was increased to 400 g on Mondays, Wednesdays, and Fridays. The sodium sulfate additions were maintained at 1,300 g on Tuesdays until day 137 when additions were terminated. Yeast extract dosing continued until day 187. High levels of TOC and COD were maintained (in excess of 10,000 mg/L) in MW-16A once the yeast extract dosing was increased to 1,200 g per week. Sulfate levels also increased over 1,000 mg/L and acted as the predominant electron acceptor in this well. Figure 2 graphically depicts these data.

The groundwater typically remained anaerobic throughout the study with Eh ranging from −165 mv to −465 mv. However, pH declined from 11.4 on day 1 to below 8.0 on day 51, apparently the result of microbial-generated buffering compounds such as organic acids and CO_2. Subsequent to termination of substrate dosing on day 187, the pH rebounded to approximately 10.8, very close to its original level. Figure 3 graphically depicts these data.

High numbers of microbes were present in the groundwater even at the low wintertime temperatures of 9°. The groundwater was turbid, had a strong anaerobic odor, and generated large quantities of gas (mostly CO_2 and some methane). The presence of microorganisms was visually confirmed by light microscopy, however microbial counts were not performed.

The concentrations of chloroethenes, chloroethanes, and chloromethanes were tracked throughout the progress of the in-field test. The concentrations of chloroethenes decreased substantially after the substrate levels reached 1,000 mg/L on day 50 (see Figure 4). The concentration of PCE decreased from a maximum level of 0.81 mg/L to less than the detection limit of 0.05 mg/L, a

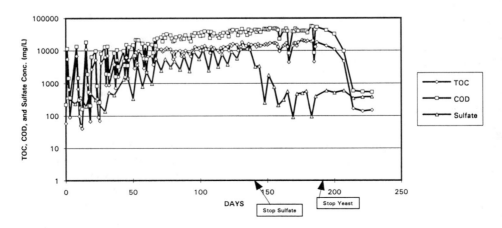

FIGURE 2. TOC, COD, and sulfate concentrations at MW-16A formation feeding.

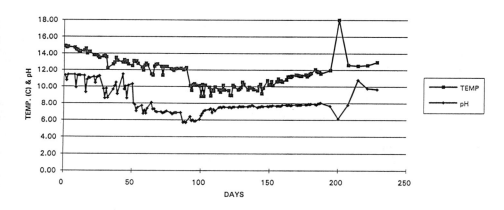

FIGURE 3. MW-16A temperature and pH.

reduction in excess of 93%. The concentrations of TCE and DCE likewise showed a substantial decrease in excess of 97%. No increase of typical daughter products from the anaerobic reductive dechlorination pathway was observed. Vinyl chloride (VC) was not detected in any of the groundwater samples analyzed. Additional analysis of well headspace gas was conducted and indicated no detectable VOCs (detection limit 0.12 ppm to 0.48 ppm).

Furthermore, the in situ concentrations of chloromethanes over time (see Figure 6) generally followed the historical pre-test patterns. Volatilization of chloromethanes was not evident from these concentration patterns. Subsequent to the cessation of substrate additions, the PCE, TCE, and DCE began to increase and approach pretest levels.

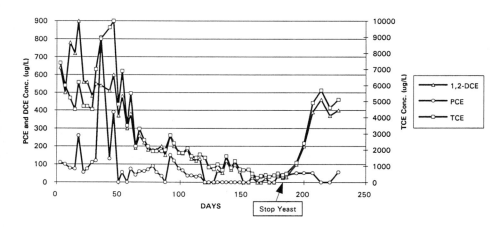

FIGURE 4. Chloroethene concentrations at MW-16A, formation feeding.

Chloroethanes were also biodegraded (see Figure 5). Concentrations of 1,1,1-TCA, and 1,1-DCA were reduced to below detection limits with approximately 97% removal. Chloroethane was not detected during the test period.

Chloroform (CF) and dichloromethane (DCM) were the only chloromethanes detected during the test (see Figure 6). These compounds behaved differently from the chloroethenes and chloroethanes. CF concentrations increased between days 57 and 194, then fell sharply after day 228 (approximately a 70% removal). The concentration of dichloromethane fluctuated during the test period and showed no discernible trend.

Ethene was detected in the groundwater samples from MW-16A. Ethene concentrations declined from 0.04 mg/L on day three to less than 0.001 mg/L and then rebounded to 0.021 mg/L on day 228. Ethane was also detected and followed a similar pattern. Ethane concentrations declined from 0.013 mg/L to less than 0.002 mg/L.

DISCUSSION

The hydrogeology in the area of MW-16A favored the maintenance of elevated levels of substrate and sulfate to enhance anaerobic reductive dechlorination. The substrate and sulfate levels were maintained at a level expected to be well in excess of that needed to promote reductive dechlorination. Total phosphate levels remained high throughout the test and approached 100 mg/L. The formation was not phosphate limited.

The microbes in MW-16A responded to the yeast extract additions with greater than 94% reduction from peak concentrations of PCE, TCE, and DCE. Similarly, reductions of peak concentrations of CF, 1,1,1-TCA, and 1,1-DCA were also substantial 94%, 91%, and 73% respectively. The low levels of chlorinated

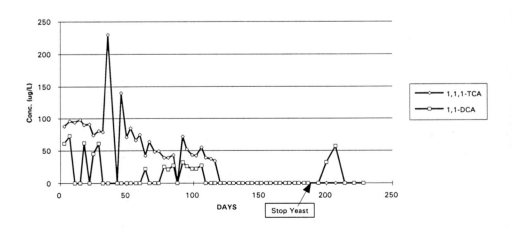

FIGURE 5. Chloroethane concentrations at MW-16A, formation feeding.

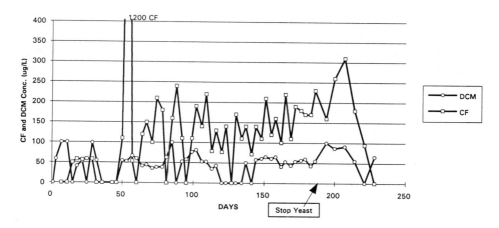

FIGURE 6. Chloromethane concentrations at MW-16A, formation feeding.

compounds found between day 158 to 187 may be the result of movement of contaminated groundwater into the well from upgradient. As a cross check on the potential fate of the chlorinated hydrocarbons, a headspace sample was collected and analyzed from well MW-16A. No detectable levels of these compounds were observed. Moreover, no discernible volatilization of chloromethanes was evident during the test sequence. Therefore, preferential volatilization of the other chlorinated hydrocarbon co-contaminants during the test is not viewed as a plausible mechanism for the noted removals and biodegradation is considered likely.

Typically, under anaerobic reductive dechlorination, daughter products such as TCE, DCE, VC, and ethene are expected to increase in proportion to the decrease of the parent compound, PCE. This sequence was not observed and leads to the supposition that another degradation pathway may have been involved in the test zone or the kinetics were so fast that these daughter products were not detected. Additional laboratory studies are planned to elucidate this mechanism and/or kinetics.

Subsequent to the termination of substrate addition on day 187, the chlorinated hydrocarbon concentrations rebounded and approached pretest levels. This is further evidence that microbial degradation was responsible for the noted removals of chlorinated aliphatic hydrocarbons.

The microbes were active even at the low wintertime temperatures encountered in the groundwater and the alkaline pH conditions. A very high microbial biomass was produced, which appeared to be fermentative based on the high levels of CO_2 evolved. Moreover, after sulfate additions ceased at day 137, there was a significant decrease in TCE levels (1.6 mg/L to 0.28 mg/L), which suggests that high sulfate levels may interfere with the electron flow from the dechlorination reaction and may actually shunt electrons to sulfate reduction.

Lastly, the microbes reduced in situ pH levels from 11.4 to less than 8.0 during the test sequence. This was strictly a microbial-mediated reaction and was accomplished without the addition of exogenous buffering agents. Subsequent to termination of the substrate additions, the pH increased to close to pretest levels.

REFERENCES

Beeman, R.E. 1994. "In Situ Biodegradation of Groundwater Contaminants," U.S. Patent 5,277,815.
Beeman, R. E., J. E. Howell, S. H. Shoemaker, E. A. Salazar, and J. R. Buttram. 1994. "A Field Evaluation of In Situ Microbial Reductive Dehalogenation by the Biotransformation of Chlorinated Ethenes." In R. E. Hinchee, A. Leeson, L. Simprini, and S. K. Ong (Eds.), *Bioremediation of Chlorinated and Polycyclic Aromatic Hydrocarbon Compounds.* Lewis Publishers, Boca Raton, FL.

AUTHOR LIST

Albrechtsen, Hans-Jørgen
Technical Univ. of Denmark
Inst. of Envir. Science & Engrg.
Groundwater Research Centre
Building 15
DK-2800 Lyngby
DENMARK

Alcantar, Celia M.
University of California, Berkeley
468 McAuley Street
230 Hesse Hall
Berkeley, CA 94720 USA

Baker, Katherine H.
Environ. Microbiology Assoc., Inc.
2001 North Front Street
Building 1, Suite 217
Harrisburg, PA 17102 USA

Barker, Gary W.
Amoco
Mail Code 19.176
501 Westlake Park Blvd.
P.O. Box 3092
Houston, TX 77253-3092 USA

Barker, James F.
University of Waterloo
Waterloo Center for Groundwater
    Research
Waterloo, Ontario N2L 3G1
CANADA

Barlaz, Morton A.
North Carolina State University
Dept. of Civil Engineering
P.O. Box 7908
Mann Hall
Raleigh, NC 27695-7908 USA

Bartlett, Craig L.
DuPont Co.
Environ. Remediation Srvcs.
300 Bellevue Parkway, Suite 390
Wilmington, DE 19809-3722 USA

Beckman, Michael A.
B.P. Barber and Associates, Inc.
5300 International Blvd., Suite 100
North Charlestown, SC 29418 USA

Benson, Leigh A.
Parsons Engineering Science, Inc.
1700 Broadway, Suite 900
Denver, CO 80290 USA

Bjerg, Poul L.
Technical University of Denmark
Dept. of Environmental Engineering
Building 115
DK-2800 Lyngby
DENMARK

Borden, Robert C.
North Carolina State University
Dept. of Civil Engineering
P.O. Box 7808
Mann Hall
Raleigh, NC 27695-7908 USA

Bouwer, Edward J.
The Johns Hopkins University
Dept. of Geog. & Environ. Engrg.
313 Ames Hall
3400 North Charles Street
Baltimore, MD 21218-2686 USA

Brown, David R.
Amoco Production Co.
P.O. Box 800
Denver, CO 80201 USA

Brown, Richard A.
Groundwater Technology, Inc.
310 Horizon Center Drive
Trenton, NJ 08691 USA

Buchanan, Jr., Ronald J.
DuPont Co.
Environmental Remediation Srvcs.
300 Bellevue Parkway, Suite 390
Wilmington, DE 19809 USA

Buscheck, Timothy E.
Chevron Research & Tech. Co.
1003 West Cutting Blvd.
Richmond, CA 94804-0054 USA

Butler, William A.
DuPont Co.
Environ. Remediation Srvcs.
300 Bellevue Parkway, Suite 390
Wilmington, DE 19809-3722 USA

Caron, Denise
U.S. Air Force
AFBCA/OL-C
Building 321
George AFB, CA 92394-5000 USA

Chiang, Chen Yu
Shell Development Co.
P.O. Box 1380
Houston, TX 77251-1380 USA

Christensen, Thomas H.
Technical University of Denmark
Inst. of Environmental Engineering
Building 115
DK-2800 Lyngby
DENMARK

Corgan, John M.
Amoco Production Co.
4502 East 41st
P.O. Box 3385
Tulsa, OK 74102 USA

Cox, Evan E.
Beak Consultants Ltd.
42 Arrow Road
Guelph, Ontario N1K 1S6
CANADA

Davis, Gregory B.
CSIRO Division of Water Resources
Perth Laboratory
Private Bag PO Wembley
Perth, Western Australia 6014
AUSTRALIA

Dortch, Ira J.
Shell Development Co.
P.O. Box 1380
Houston, TX 77251-1380 USA

Doyle, Greg
IT Corporation
1425 S. Victoria Court, Suite A
San Bernadino, CA 92408-2923 USA

Durant, Neal D.
The Johns Hopkins University
Dept. of Geog. & Environ. Engrg.
313 Ames Hall
3400 North Charles Street
Baltimore, MD 21218 USA

Edwards, Elizabeth
Beak Consultants Ltd.
42 Arrow Road
Guelph, Ontario N1K 1S6
CANADA

Ellis, David E.
DuPont Co.
Engineering Department
300 Bellevue Parkway, Suite 390
Wilmington, DE 19809-3722 USA

Fisher, George A.
DuPont Co.
301 Glasgow Business Center
P.O. Box 6101
Newark, DE 19714-6101 USA

Fisher, J. Berton
Amoco
4502 East 41st Street
P.O. Box 3385
Tulsa, OK 74102 USA

Gannon, John T.
DuPont Co.
301 Glasgow Business Center
P.O. Box 6101
Newark, DE 19714-6101 USA

Ginn, Jon S.
Utah State University
Utah Water Research Laboratory
Civil & Environ. Engineering Dept.
1600 East Canyon Road
Logan, UT 84322-8200 USA

Guest, Peter R.
Parsons Engineering Science, Inc.
1700 Broadway, Suite 900
Denver, CO 80290 USA

Hamilton, Kimberly A.
University of Waterloo
Waterloo Centre for Groundwater
    Research
Waterloo, Ontario N2L 3G1
CANADA

Hansen, Jerry E.
U.S. Air Force
HQ AFCEE/ERT
8001 Arnold Drive
Brooks AFB, TX 78235-5357 USA

Hare, Paul W.
General Electric Company
1 Computer Drive South
Albany, NY 12205 USA

Heron, Gorm
Technical University of Denmark
Inst. of Environ. Science & Engrg.
Groundwater Research Centre
Building 115
DK-2800 Lyngby
DENMARK

Herson, Diane S.
University of Delaware
Biology Department
Newark, DE 19716 USA

Hicks, Patrick M.
Groundwater Technology, Inc.
3110 Cherry Palm Drive, Suite 390
Tampa, FL 33619 USA

Hicks, Ronald J.
Groundwater Technology, Inc.
4080 Pike Lane
Concord, CA 94520 USA

Holmes, Marty
DuPont Co.
301 Glasgow Business Center
P.O. Box 6101
Newark, DE 19714-6101 USA

Hunt, Melody J.
North Carolina State University
Dept. of Civil Engineering
208 Mann Hall
P.O. Box 7908
Raleigh, NC 27695 USA

Jonkers, Constance A. A.
The Johns Hopkins University
Dept. of Geog. & Environ. Engrg.
313 Ames Hall
3400 North Charles Street
Baltimore, MD 21218 USA

Kampbell, Donald H.
U.S. Environ. Protection Agency
RS Kerr Environ. Research Lab
919 Research Drive
P.O. Box 1198
Ada, OK 74820 USA

King, Mark W. G.
University of Waterloo
Waterloo Centre for Groundwater
    Research
Waterloo, Ontario N2L 3G1
CANADA

Leahy, Maureen C.
Groundwater Technology, Inc.
Kennedy Bus. Park 2
431 F. Hayden Station Road
Windsor, CT 06095 USA

Lee, Michael D.
DuPont Co.
Environ. Remediation Srvcs.
Glasgow Building 300
P.O. Box 6101
Newark, DE 19714-6101 USA

Lehmicke, Leo
Beak Consultants
12931 N.E. 126th Place
Kirkland, WA 98034-7716 USA

Ludvigsen, Liselotte
Technical University of Denmark
Inst. of Environ. Science & Engrg.
Groundwater Research Centre
Building 115
DK-2800 Lyngby
DENMARK

Major, David W.
Beak Consultants Ltd.
42 Arrow Road
Guelph, Ontario N1K 1S6
CANADA

Mazierski, Paul F.
DuPont Co.
Environmental Remediation Srvcs.
Buffalo Avenue & 26th St., Bldg. 35
Niagara Falls, NY 14302 USA

McAllister, Paul M.
Shell Development Co.
3333 Highway 6S
P.O. Box 1380
Houston, TX 77251-1380 USA

Miller, Ross N.
U.S. Air Force
8001 Arnold Drive, Building 642
Brooks AFB, TX 78235-5357 USA

Mosbæk, Hans
Technical University of Denmark
Inst. of Environ. Science & Engrg.
Building 115
DK-2800 Lyngby
DENMARK

Murarka, Ishwar P.
Electric Power Research Institute
Palo Alto, CA 94303 USA

Newell, Charles J.
Groundwater Services, Inc.
5252 Westchester, Suite 270
Houston, TX 77005 USA

Odom, J. Martin
DuPont Co.
301 Glasgow Business Center
P.O. Box 6101
Newark, DE 19714-6101 USA

Perkins, Richard E.
DuPont Co.
Central Research & Development
301 Glasgow Business Center
P.O. Box 6101
Newark, DE 19714-6101 USA

Power, Terry R.
CSIRO Division of Water Resources
Private Bag, PO Wembley
Perth, Western Australia 6014
AUSTRALIA

Rainsberger, Tony J.
Parsons Engineering Science, Inc.
1700 Broadway, Suite 900
Denver, CO 80290 USA

Raterman, Kevin T.
Amoco Production Co.
P.O. Box 3385
4502 East 41st Street
Tulsa, OK 74102-3385 USA

Rifai, Hanadi S.
Rice University
Energy and Environ. Systems Inst.
P.O. Box 1892, MS 316
Houston, TX 77251-1892 USA

Rounsaville, Mark
U.S. Air Force
8001 Arnold Drive, Building 642
Brooks AFB, TX 78235-5357 USA

Rügge, Kirsten
Technical University of Denmark
Inst. of Environ. Science & Engrg.
Building 115
DK-2800 Lyngby
DENMARK

Salanitro, Joseph P.
Shell Development Co.
P.O. Box 1380
Houston, TX 77251-1380 USA

Sehayek, Lily S.
DuPont Co.
Environ. Remediation Srvcs.
300 Bellevue Parkway, Suite 390
Wilmington, DE 19809-3722 USA

Sewell, Guy W.
U.S. Environ. Protection Agency
Robert S. Kerr Environmental
    Research Lab
919 Research Drive
P.O. Box 1198
Ada, OK 74820 USA

Sims, Ronald C.
Utah State University
Dept. of Civil & Environ. Engrg.
Logan, UT 84322-4110 USA

Sublette, Kerry L.
University of Tulsa
Dept. of Chemical Engineering
600 South College Avenue
Tulsa, OK 74104 USA

Swanson, Matthew A.
Parsons Engineering Science, Inc.
1700 Broadway, Suite 900
Denver, CO 80290 USA

Swindoll, C. Michael
DuPont Co.
Environ. Remediation Srvcs.
301 Glasgow Business Center
P.O. Box 6101
Newark, DE 19714-6101 USA

Taffinder, Sam
U.S. Air Force
HQ AFCEE/ERT
8001 Arnold Drive, Building 642
Brooks AFB, TX 78235-5357 USA

Toze, Simon G.
CSIRO Division of Water Resources
Private Bag, PO Wembley
Perth Western Australia 6014
AUSTRALIA

Trent, Gary L.
Amoco Production Co.
P.O. Box 3385
Tulsa, OK 74102-3385 USA

Troy, Marleen A.
Groundwater & Envir. Srvcs., Inc.
1340 Campus Parkway
P.O. Box 1750
Wall, NJ 07719 USA

Ward, C. Herb
Rice University
Energy & Environ. Systems Inst.
P.O. Box 1892 MS-316
Houston, TX 77251-1892 USA

Wiedemeier, Todd H.
Parsons Engineering Science, Inc.
1700 Broadway, Suite 900
Denver, CO 80290 USA

Williams, Patty
Shell Development Company
P.O. Box 1380
Houston, TX 77251-1380 USA

Wilson, John T.
U.S. Environ. Protection Agency
Robert S. Kerr Environmental
    Research Laboratory
919 Research Drive
P.O. Box 1198
Ada, OK 74820 USA

Wilson, Liza P.
The Johns Hopkins University
Dept. of Geog. & Environ. Engrg.
313 Ames Hall
3400 North Charles Street
Baltimore, MD 21218 USA

INDEX